D0889762
WITHDRAWN

NORTH DAKOTA
STATE UNIVERSITY

The Scientific Attitude

The Scientific Attitude

Defending Science from Denial, Fraud, and Pseudoscience

Lee McIntyre

The MIT Press
Cambridge, Massachusetts
London, England

This book was set in Stone Serif by Westchester Publishing Services. Printed and bound in the United States of America.

Library of Congress Cataloging-in-Publication Data

Names: McIntyre, Lee C., author.
Title: The scientific attitude : defending science from denial, fraud, and pseudoscience / Lee McIntyre.
Description: Cambridge, MA : The MIT Press, [2019] | Includes bibliographical references and index.
Identifiers: LCCN 2018037628 | ISBN 9780262039833 (hardcover : alk. paper)
Subjects: LCSH: Science—Social aspects. | Science—Methodology. | Pseudoscience.
Classification: LCC Q175.5 .M3955 2019 | DDC 306.4/5—dc23
LC record available at https://lccn.loc.gov/2018037628

10 9 8 7 6 5 4 3 2 1

For Louisa and James

Contents

Preface

This book has been a labor of love from the beginning and—as with any labor—it has taken a while. I remember the exact moment when I decided to become a philosopher of science, as I was reading Karl Popper's enchanting essay "Science: Conjectures and Refutations," in the fall of 1981, in one of the upper carrels in Olin Library at Wesleyan University. The issues seemed earth-shattering and the romance was obvious: here was a person who had found a way to defend one of the ideas I believed in most—that science was special. Popper made it his life's work to defend the epistemic authority of science and explain why it was superior to its imposters. How could I not want to be involved in that?

Though the issues gripped me, I never fully agreed with Popper's conclusions. I knew I'd get back to it someday but, as the professional reward system in academia seemed to favor taking somewhat smaller bites of the apple, I contented myself with spending the first decade of my career writing about the importance of laws and prediction, how to improve the methodology of the social sciences, and why we needed a philosophy of chemistry. Since then I have taken great enjoyment in expanding my reach to write philosophy for a general audience on topics such as science denial, the importance of reason, and why—especially in this day and age—even the staunchest philosophical skeptics need to defend the idea that truth matters.

But this is the book that I have always wanted to write. By taking up a topic as important as what is distinctive about science, I hope that it will be of interest to philosophers, scientists, and the general public alike.

For inspiring me to go into philosophy I would like to thank my teachers: Rich Adelstein, Howard Bernstein, and Brian Fay. Although I overlapped with him only toward the end of my college career, Joe Rouse was also an inspiration to me. In graduate school at the University of Michigan, I had the good fortune to learn the philosophy of science from Jaegwon Kim, Peter Railton, and Larry Sklar. I was not always happy in graduate school (who is?), but I look back on my education there as the foundation for all my further work.

Since then, I am grateful to have worked with some of the best in the business: Dan Little, Alex Rosenberg, Merrilee Salmon, and Eric Scerri, all of whom have taught me much through their excellent scholarship and warm colleagueship. My debt to Bob Cohen and Mike Martin—both of whom passed away in recent years—is enormous, for they gave me a home in the philosophy of science and helped me along at every step of the way. I am glad to say that the same has been true of the new director of the Center for Philosophy and History of Science at Boston University, Alisa Bokulich, as well.

For guidance and advice on some of the specific ideas contained in this book, I would like to thank Jeff Dean, Bob Lane, Helen Longino, Tony Lynch, Hugh Mellor, Rose-Mary Sargent, Jeremy Shearmur, and Brad Wray. I was lucky to have been a participant in Massimo Pigliucci and Maarten Boudry's workshop on "scientism" at CUNY in the spring of 2014, where I heard some extremely stimulating papers by Noretta Koertge, Deborah Mayo, and Tom Nickles, that inspired me to think about writing this book. Rik Peels and Jeff Kichen made pinpoint suggestions about discrete problems that were enormously helpful as well.

My good friends Andy Norman and Jon Haber have done me the honor of reading the complete manuscript of this book in draft and making many helpful suggestions. My friend Laurie Prendergast has once again done yeoman service by helping me with the proofreading and index. I would also like to acknowledge the work of five anonymous referees, whom I obviously cannot thank by name, each of whom made enormous contributions to the content of this book. It goes without saying that any remaining mistakes are mine alone.

My father unfortunately did not live to see the publication of this book, but to him and to my mother I send all my love and gratitude for always believing in me and for their support and guidance over the years. My wife Josephine, and children Louisa and James, each read this book in detail and lived with its ups and downs through many iterations. No man was ever luckier to be married to such a wonderful woman, who wants nothing more than for me to be happy in my life and in my work. I am fortunate also to have not one but *two* children who majored in philosophy and claim it as their birthright to identify any flaws in the old man's reasoning, which they have done with frightening efficiency. Indeed, my children's contributions to this book have been so great that I would like to dedicate it to them.

The team at MIT Press is without parallel. As they proved in my last book with them—and every day since—no author is ever successful alone. From copyediting to design, and marketing to editorial, it is my privilege to work with all of them. Here I would like to give special thanks to my tireless and creative publicity team and to my editor Phil Laughlin, who is analytical, succinct, practical, and hilarious all at the same time. Once again they have made it a pleasure to work with MIT Press, in what is now my fourth book with them.

My final debt is an old one, but I see it every day as I look at the framed handwritten letter that I received from Karl Popper in March 1984, in response to a letter I wrote to him as an undergraduate. Popper was brilliant, lucid, defensive, and enlightening. Although I disagree with many of his ideas in the philosophy of science, I could not have developed my own without having had his to react against and—in one of the most delightful discoveries of my career—found that in some ways he had already anticipated the scientific attitude. I never met Karl Popper, but my earliest vision of him stays with me: a young man just starting out in the winter of 1919, realizing the logic of falsification in a lightning flash, then working out its details over the course of his career. I was proud to learn that this book would be published precisely one hundred years after Popper's discovery. It is small tribute to a man who inspired my own and so many other careers in the philosophy of science.

Introduction

We live in extraordinary times for the understanding of science. In May 2010, the prestigious journal *Science* published a letter signed by 255 members of the US National Academy of Sciences. It began "We are deeply disturbed by the recent escalation of political assaults on scientists in general and on climate scientists in particular. All citizens should understand some basic scientific facts. There is always some uncertainty associated with scientific conclusions; science never absolutely proves anything."[1]

But how many laypeople understand what this means and recognize it as a strength rather than a weakness of scientific reasoning? And of course there are always those who are willing to exploit any uncertainty for their own political purposes. "We don't know what's causing climate change, and the idea of spending trillions and trillions of dollars to try and reduce CO_2 emissions is not the right course for us," said US presidential candidate Mitt Romney in 2011.[2] In the following election cycle, in an interview in which he questioned whether there was really any good evidence for global warming, US Senator Ted Cruz said, "Any good scientist questions all science. If you show me a scientist that stops questioning science, I'll show you someone who isn't a scientist."[3] Scarcely a year later, newly elected President Donald Trump said that he wanted to eliminate all climate change research done by NASA, in an effort to crack down on "politicized science." This would mean an irreparable loss for climate monitoring, not only for the United States but for all researchers the world over who depend on NASA's legendary satellite-driven data collection about temperature, ice, clouds, and other phenomena. As one scientist from the National Center for Atmospheric Research put it, "[this] could put us back in the 'dark ages' of almost the pre-satellite era."[4]

The attack on science has now gotten so bad that on April 22, 2017, there was a "March for Science" in six hundred cities around the world. At the one in Boston, Massachusetts, I saw signs that said "Keep Calm and Think Critically," "Extremely Mad Scientist," "No Science, No Twitter," "I Love Reality," "It's So Severe, The Nerds Are Here," and "I Could Be in the Lab Right Now." It takes a lot to get scientists out of their labs and onto the streets, but what else were they supposed to do? The issue of what's special about science is no longer purely academic. If we cannot do a better job of defending science—of saying how it works and why its findings have a privileged claim to believability—we will be at the mercy of those who would thoughtlessly reject it.

The aim of this book is to understand what is distinctive about science. Of course, some might say that we don't need to do this because it has already been done; that the problem is in *communicating* what is special about science, not understanding it. *Don't we already know what's special about science by looking at the work of scientists? And, if not, isn't there plenty of work done by other philosophers of science that can answer this question?* I wish this were true, but the fact is that most scientists tend to be "naive realists" who accept their findings as revealing something true (or close to true) about nature and spend little time considering the philosophical or methodological issues behind science as a whole. Those rare scientists who do venture into philosophy usually stumble over something that philosophers have already discovered or end up blasting the whole enterprise as irrelevant because the point—they argue—is not to reflect upon science but to *do* it.[5] Yet that's just the problem. For all the success of those who have *done* science, why do so many still feel at a loss to respond with anything other than inarticulate name-calling to those who say that science is "just another ideology" or that we "need more evidence" on climate change? There has to be a better way. Better both to justify the science that has already been done, but also to lay the groundwork for good science to grow elsewhere in the future. But first we must understand what is so special about science as a way of knowing. And for this, many have turned to the philosophy of science.

The foundation of the philosophy of science since its inception has been the idea that it can make a unique contribution by providing a "rational reconstruction" of the process of science, in answer to the question of why science works as well as it does (and why its claims are justified).[6] There is a good deal of debate, however, over the best means for doing this and whether it is even a worthy aim. The idea that we can transplant science into other fields by understanding what is most distinctive about it has gotten something of a bad reputation over the years. This notoriety has come from those who have claimed that there is a "scientific method"—or some other firm criterion of demarcation between science and nonscience—such that if we could just apply the standard rigorously enough, good science would bloom as a result. Such claims are made worse by those who embrace the spirit of proselytizing and engage in what has been called "scientism," whereby they now have a hammer and every other field in the universe of inquiry looks like a nail. But there is a problem: nearly everyone in the philosophy of science these days admits that there is no such thing as scientific method, that trying to come up with a criterion of demarcation is old-fashioned, and that scientism is dangerous.[7] Along the way, most have also largely given up on the idea that prescription lies at the heart of the philosophy of science.

Karl Popper's 1934 model of science *Logik der Forschung*, translated into English in 1959 by Popper himself as *The Logic of Scientific Discovery*, focuses heavily on the idea that there is a reliable method for demarcating science from nonscience, but that there is no such thing as a "scientific method." Popper champions the idea that science uses "falsifiable" theories—ones that are capable at least in principle of being proven wrong by some evidence—as the dividing line. Although this model has several logical and methodological virtues, it has also proven problematic for many philosophers of science, who complain that it is too idealized and focuses too heavily on the "greatest moments" of science, like the transition from the Newtonian to the Einsteinian model in physics, and that most science does not actually work like this.[8]

Another account was offered by Thomas Kuhn in 1962, in his famous book *The Structure of Scientific Revolutions*. Here the focus is on how some scientific theories replace others through paradigm shifts, where the scientific consensus changes radically as a result of problems that have built up with the old theory, and the field switches seemingly overnight to a new one. The problem here, however, is not only the familiar complaint that most science does not actually work like this (for instance, the transition from the Ptolemaic Earth–centered model to the Copernican Sun–centered model in astronomy)—which Kuhn freely admits when he talks about the ubiquity of "normal science"—but that even where it does, this is not a completely "rational" process. Although Kuhn insists on the key role of evidence in paradigm change, once we have opened the door to "subjective" or social factors in interpreting this evidence, there seems to be no "method" to follow.[9] This not only presents a problem for showing that scientific claims are justifiable, but also forestalls delineating a roadmap for other sciences.

Still further models of scientific change have been proposed by Imre Lakatos, Paul Feyerabend, Larry Laudan, and the "social constructivists," each of whom drain a little more water from the pool that allows us to say that science is "special" and that other fields of inquiry would do well to follow its example.[10] So what to do? Just pick one of the existing accounts? But this is not possible. For one thing, they are largely incompatible with one another; each describes a different piece of the "blind man's elephant," so we are still missing a comprehensive picture of what science is like. Another problem is that these models seem to succeed only to the extent that they leave something behind, namely the motivation that if we finally understood science we could provide a standard for other fields to become more scientific.

If all of the best accounts fail, perhaps there is a more general weakness in this whole approach? Although some may be loath to identify it as a weakness, it seems at least a drawback that the philosophy of science has spent so much of its time focusing on the "successes" of science and has not had very much to say about its failures. In fact, the lessons of failure to live up to the scientific standard are as revealing about what science is as the example of those fields that have achieved it. There is nothing wrong in principle with exploring what is distinctive about science by looking at its accomplishments, but this has led to some mischief.

First, while it would be comforting to imagine science as a long series of steps toward the truth—with its failures due only to the wrongheaded and ignorant—this view of science is belied by its history, which is littered with theories that were scientific but just turned out to be wrong. Both Popper and Kuhn have done much to show how science is strengthened by an uncompromising focus on explanatory "fit" between theory and evidence, but it is all too easy for others to look back and pretend that this was all inevitable and that the arc of science bends irrevocably toward a single (true) explanation of reality.

Second, the relentless focus on explaining science through its successes has meant that most of the "victories" that philosophers can turn to for their models have come from the natural sciences. Specifically, we have been forced to draw most of our conclusions about what makes science special from the history of physics and astronomy. But this is a bit like drawing targets around where the darts have landed. And does this mean that in attempting to be scientific, other fields should try to emulate physics? Thinking that the answer to this question is an unqualified yes has done a great disservice to other fields, some of which are solidly empirical but quite different in subject from the physical sciences. Remember that one important part of the mission of the philosophy of science is to understand what is distinctive about science *so that we can grow it elsewhere*. But where does this leave fields like the social sciences, which until quite recently have been underserved by most of the explanatory models in the philosophy of science?

Popper famously argued that the social sciences could not be sciences because of the "open systems" problem created by the effect of free will and consciousness on human decision making. In natural science, he claimed, we use falsifiable theories, but this path is not open to the social sciences.[11] Similarly, Kuhn, for all his fans in the social sciences (who felt that they may finally have a target they could hit), also tried to distance himself from the messy study of human behavior, by insisting that his model was applicable only to the natural sciences, and that he was not providing any advice to the social sciences. Add to this the problem of what to do about some of the other "special" (i.e., nonphysical) sciences—like biology or even chemistry—and we have a full-fledged crisis on our hands in defending a view of science that is separate from reduction to physics. What to do about the claim that there are epistemically autonomous concepts in chemistry (such as *transparency* or *smell*)—just as there are in sociology (such as *alienation* or *anomie*)—that cannot be explained at the physical level of description? If our model of successful science is physics, will even chemistry make the cut? From a certain perspective, *most* of those fields that are either scientific, or wish to become so, do not fit the models of philosophy of science that have been drawn from the history of physics and astronomy, and so could be considered "special sciences." Have we no advice, or justification, to offer them?

Finally, what to say about those fields that make a claim to being scientific, but just do not measure up (such as "intelligent design theory" or denialism about climate change)? Or of those instances where scientists have betrayed their creed and committed fraud (such as Andrew Wakefield's work purporting to show a link between vaccines and autism)? Can we learn anything from them? I maintain that if we are truly interested in what is special about science, there is much to learn from those who have forsaken it. What is the proponent of intelligent design theory *not* doing that genuine scientists *should* do (and in fact generally succeed in doing)? Why are climate change deniers unjustified in their high standards of "skepticism"? And why is it forbidden for scientists to rig their data, cherry pick their sample sets, and otherwise try to fit the data to their theory, if they want to succeed in scientific explanation?[12] It may seem obvious to those who defend and care about science that all of the above have committed a mortal sin against scientific principles, but shouldn't this help us in articulating the nature of those principles?

In this book, I propose to take a very different approach from my predecessors, by embracing not only the idea that there is something distinctive about science, but that the proper way to understand it is to eschew the exclusive focus on the successes of natural science. Here I plan to focus on those fields that have failed to be scientific, as well as those (like the social sciences) that might wish to become more so. It is one thing to discern what is distinctive about science by examining the transition from Newton to Einstein; it is another to get our boots muddy in the questions of scientific fraud, pseudoscience, denialism, and the social sciences.

Why bother? Because I think that to truly understand both the power and the fragility of science we must look not just at those fields that are already scientific, but also at those that are trying (and perhaps failing) to live up to the standard of science. We can learn a lot about what is special about science by looking at the special sciences. And we should be prepared to answer the challenge of those who want to know—if science is so credible—why it does not always provide the right answer (even in the natural sciences) and sometimes fails. If we can do this, we will not only understand what is distinctive about science, we will have the tools necessary to emulate its approach in other empirical fields too.

But there is another problem: we cannot pretend, these days, that the conclusions of science are going to be accepted just because they are rational and justified. Climate change skeptics insist that we need more evidence to prove global warming. Vaccine resisters maintain that there is a conspiracy to deny the truth about autism. What should we do about the problem of those who would simply reject the results of science? We may be tempted to dismiss these people as irrational, but we do so at our peril.

If we cannot provide a good account of why scientific explanations have a superior claim to believability, why should they accept them? It's not just that if we don't understand science we cannot cultivate it elsewhere; we cannot even defend science where it is working.

In short, I think that many of those who have written about science have mishandled the claim that science is special because they have not said enough about the failures of natural science, the potential for the social sciences, and the drawbacks of those fields that seek the mantle of science without embracing its ethos. This has led to failure to emulate science by those fields that would like to do so, and also to the irrational rejection of scientific conclusions by those who are motivated by their ideologies to think that their own views are just as good.

So what *is* distinctive about science? As I hope to show, what is most notable is the *scientific attitude* toward empirical evidence, which is as hard to define as it is crucial. To do science we must be willing to embrace a mindset that tells us that our prior beliefs, ideologies, and wishes do not matter in deciding what can pass the test of comparison with the evidence. This is no easy thing to mark off with a criterion of demarcation—neither does it pretend to be a proxy for "scientific method"—but I argue that it is essential to engaging in (and understanding) science. This is something that can be emulated by the social sciences and also helps to explain what is *not* scientific about intelligent design theory, the emptiness of denialism by those who wish to reject the evidence for climate change, and the folly of other conspiracy theories that purport to succeed where science is restrained by bona fide skepticism. At its heart, what is distinctive about science is that it *cares about evidence* and is *willing to change its theories on the basis of evidence*. It is not the subject or method of inquiry but the values and behavior of those who engage in it that makes science special. Yet this is a surprisingly complex thing to unravel, both in the history of the past successes of science and also in a program for how to make other fields more scientific in the future.

In the chapters to follow, I will show how the scientific attitude helps us with three main tasks: understanding science (chapters 1 through 6), defending science (chapters 7 and 8), and growing science (chapters 9 and 10). When done right, the philosophy of science is not just descriptive or even explanatory, but prescriptive. It helps to explain not just why science has been so successful in the past, but why evidential and experimental methods have so much potential value for other empirical fields in the future. It should also help us to communicate more clearly to those who do not—or will not—understand what is distinctive about science, why the claims of pseudoscience and denialism fall so far short of its epistemic standards, and why scientific explanations are superior. For decades, philosophers of science have sought to understand what is special about science by focusing on the past successes of the physical sciences. My approach, instead, is to turn this on its head: if you really want to understand why science is so special, you must look beyond the victories of natural science, and focus too on those fields that are not—and may never become—sciences.

1 Scientific Method and the Problem of Demarcation

If there is one thing that most people think is special about science, it is that it follows a distinctive "scientific method." If there is one thing that the majority of philosophers of science agree on, it is the idea that there is no such thing as "scientific method."

If you are one of those people who has saved his or her textbooks in astronomy, physics, chemistry or biology, I invite you to go find one now and open it to the first page. This is typically the page that is never assigned by the professor and never read by the students, but it is nonetheless *de rigeur*, because it purports to provide the basis for why the rest of the claims in the book are to be believed. Often this page gives an account of "scientific method." There are different renderings, but here is a simple version of the classic so-called five step method:

(1) Observe.

(2) Hypothesize.

(3) Predict.

(4) Test.

(5) Analyze results, revise hypothesis, and start again.[1]

Is this how scientific discoveries are in fact made? Few would claim this. The way that scientific theories are *produced* is an often messy process that involves serendipity, failure, blind alleys, heartache, dogged determination, and the occasional stroke of luck. But that is not what is supposed to be special about science. The quirky way that scientists sometimes get their ideas is the stuff of legend. One thinks of August Kekulé in front of the fireplace, dreaming of a snake swallowing its tail, leading to the benzene ring, or Leo Szilard stepping off the curb as the traffic light changed from red to green, realizing in that moment that it was possible to split the atom.[2] Inspiration in science, as in art, can come from a diversity of sources. Yet many hold that the results of science have a special claim to be believed because of the distinctive way that they can be rationally reconstructed *after the fact*. Thus it is not the way that scientific theories are *found* that gives them such great credibility, it is the process by which they can be *logically justified*.

Science textbooks provide a cleaned-up version of history. They give us the result of many centuries of scientific conflict and make us feel that the process inevitably led to our present enlightened understanding. Historians of science know this to be inaccurate, but it remains immensely popular, because of the great convenience that this account of "scientific method" provides in supporting not only the claim that the content of science is especially credible but also the idea that the process of scientific explanation can be emulated by other disciplines that wish to make their own empirical discoveries.[3]

Yet even if the classic five-step method proves too simple to get the job done, there are other ways that philosophers have sought to characterize what is distinctive about science, and some of these focus on methodology. Here it is important not to get confused. The claim that there is no universal one-size-fits-all "scientific method"—where we put in sensory observations at one end and get scientific knowledge at the other—does not contradict the idea that there could be some unique methodological feature of science. To say that there is no recipe or formula for producing scientific theories is a very different thing than to claim that scientists have *no methods whatsoever.* This is to say that even if most philosophers of science are willing to reject the idea of a simple "scientific method," many still think there is enormous benefit to analyzing the methodological differences between science and nonscience, in search of a way to justify the epistemic authority of those scientific theories that have already been discovered.

The Relevance of the Problem of Demarcation

One benefit of focusing on the methodology of science is that it purports to provide a way to demarcate between what is science and what is not. This is the so-called *problem of demarcation*, and it has been of enormous concern to the philosophy of science at least since the time of Karl Popper at the beginning of the twentieth century. In his essay "The Demise of the Demarcation

Problem," Larry Laudan claims that the problem of demarcation goes all the way back to Aristotle—who sought to differentiate between knowledge and opinion—and surfaced again in the era of Galileo and Newton—who pushed science into the modern era by using empirical methods to understand how nature worked. By the beginning of the nineteenth century, Laudan asserts, Auguste Comte and others began to hone in on the claim that what was distinctive about science was its "method," even if there was as yet no widespread agreement about what that actually was.[4] By the beginning of the twentieth century, philosophers were ready to sharpen this analysis and attempt to solve the problem of demarcation by creating a strict "criterion of demarcation" that could differentiate science from nonscience.

The Logical Positivists tried to do this on the basis of the allegedly special "meaning" of scientific statements. In contrast to other kinds of assertions, scientific statements were accepted as making a difference to our experience in the world, which meant that they must in some way be verifiable through sensory data. If scientists said that the planet Venus went through phases, we had better be able to see that through a telescope. Statements that could not do this (other than those used in logic, which is deductively valid and so already on firm ground) were branded "cognitively meaningless" because they were unverifiable; they were dismissed as nonsense and unworthy of our time, because there was no procedure for determining if they were right or wrong. If a statement about the world purported to be true, the positivists claimed, it must be verifiable by experience. If not, then instead of being scientific it was just "metaphysics" (which was the pejorative term they used to cover huge swaths of knowledge including religion, ethics, aesthetics, and the vast majority of philosophy). To pull off such a hard and fast distinction, though, the Logical Positivists needed to come up with a "verification criterion" by which one could sort meaningful from meaningless statements. And this, owing to technical reasons that ultimately came down to the problem that they couldn't get the sorting right, led to their undoing.[5]

The problem of demarcation was then taken up by perhaps its greatest champion, Karl Popper. Popper understood—even before Logical Positivism had formally failed—that there were problems with pursuing the verification of scientific statements. The positivists had based science on inductive inference, which undercut the idea that one could prove any empirical statement to be true. David Hume's famous problem of induction prevented the sort of logical certainty that the positivists coveted for scientific statements.[6] Even if they could not be proven, however, weren't scientific assertions nonetheless uniquely meaningful, given the fact that in principle they *could* be verified? Popper thought not, and regarded the pursuit of "meaning" as another mistake in the positivists' approach. What made science special, he felt, was not its meaning but its method. Popper thus set out, in the winter of 1919, to try to solve the problem of demarcation in another way—by renouncing both verification and meaning, focusing instead on what he called the "falsifiability" of scientific theories: the idea that they must be capable of being ruled out by some possible experience.

What concerned Popper was the difference between statements like those in astrology—that seemed compatible with any evidence—and those of science, that take some risk of being wrong. When an astrologer produces a personalized horoscope that says "You are sometimes insecure about your achievements and feel like an imposter" this can feel like a stunning insight into your inner-most thoughts, until you realize that the same horoscope is used for all clients. Contrast this with what happens in science. When a scientist makes a prediction, it comes with an understanding that if her theory is correct you will see what was predicted. And if you do *not* see that result, then the theory must be flawed.

Popper used this sort of contrast to think about the possible methodological difference between science and nonscience. He was searching for a way to forgo the impossibly high standard which said that scientific statements must always be *proven* by their evidence, but would still allow evidence to count. And then it hit him. If the Logical Positivists and others were searching for a way to differentiate science from nonscience—but were blocked from being able to say that scientific statements were *verifiable* because of Hume's problem of induction—why not instead follow the path of deductive certainty that was already enjoyed by logic?

Those who have studied formal logic know that the simplest and most famous deductively valid inference is *modus ponens*, which says "If A, then B. And A. Therefore B." No problem here. No need to check to see whether it "makes a difference to our experience." Deductive arguments are and always will be valid because the truth of the premises is sufficient to guarantee the truth of the conclusion; if the premises are true, the conclusion will be also. This is to say that the truth of the conclusion cannot contain any information that is not already contained in the premises. Consider the following valid argument:

If someone was born between 1945 and 1991, then they have Strontium-90 in their bones.

Adam was born in 1963.

Therefore, Adam has Strontium-90 in his bones.[7]

The problem with scientific statements, however, is that they don't seem to follow this form. For hundreds of years before Popper, they were accepted as being inductive, which meant that the reasoning looked more like "If A, then B. And B. Therefore A." For example:

If someone was born between 1945 and 1991, then they have Strontium-90 in their bones.

Eve has Strontium-90 in her bones.

Therefore, Eve was born between 1945 and 1991.

Obviously, this kind of argument is not deductively valid. The fact that Eve has Strontium-90 in her bones is no guarantee that she was born between 1945 and 1991. Eve might, for example, have grown up near a nuclear reactor in Pennsylvania in the late 1990s, where it was found that Strontium-90 was present as a result of environmental contamination. Here the form of the argument does not guarantee that if there the premises are true, the conclusion will be true. With inductive arguments, the conclusion contains information that goes beyond what is contained in the premises. This means that we will have to engage in some actual experience to see if the conclusion is true. But isn't this how we do science? Indeed, when we are engaged in reasoning about empirical matters, we often seek to go beyond our firsthand experiences and draw inferences about those situations that are similar to them. Even though our experience may be limited, we look for patterns within it and hope to be able to extrapolate them outward.

Suppose we are interested in a straightforward empirical issue such as the color of swans. We've seen a lot of swans in our life and they have all been white. We may therefore feel justified in making the assertion "All swans are white." Is this true? We've made our observations and have come up with a hypothesis, but now it is time to test it. So we make a prediction that from now on *every* swan we see will be white. Here is where it gets interesting. Suppose that this prediction turns out to be fulfilled. We may live our whole lives in North America, and, as it turns out, every single swan we ever see is white. Does this prove the truth of our assertion? No. It is still possible that someday if we go to Australia (or just open Google), we will see a black swan.

When we are trying to discover empirical truths about the world, we are hampered by the fact that our experience is always finite. No matter how long we live, we cannot possibly sample all of the swans who have lived or ever will live. So we can never be certain. If we wish to make blanket statements about the world—sometimes instantiated in scientific laws—we face the in principle worry that some future piece of evidence may come along to refute us. This is because the form of the argument that we are using here is inductive, and inductive inferences are not deductively valid. There is just no way to be certain that the rest of the world will conform to our limited experience.

Science nonetheless works pretty well. Although it may not guarantee the truth of our assertions, we are at least gathering evidence that is relevant to the warrant for our beliefs. And shouldn't this increase the likelihood that our general statements are true?[8] But why settle for this? Popper was bothered by the inductive form of inference used by positivists and others as the basis for science. But if that is its logical foundation, how can we possibly demarcate science from nonscience? To admit that "we could be wrong" doesn't sound like much of a distinction. Popper sought something stronger. He wanted a *logical* basis for the uniqueness of science.

Popper didn't have to look far. The inductive argument that we used above has a name—"affirming the consequent"—and it is a well-known fallacy of deductive logic. But there are other, better forms of argument, and one of the most powerful—*modus tollens*—is deductively valid. It works like this. "If A, then B. And not B. Therefore, not A."

If someone was born between 1945 and 1991, then they have Strontium-90 in their bones.

Gabriel does not have Strontium-90 in his bones.

Therefore, Gabriel was not born between 1945 and 1991.[9]

This was Popper's insight: this, he felt, was the logical basis for scientific inference. Just because science seeks to learn from empirical facts about the world, this does not mean that it is doomed to the problems of inductive inference. For if we look at the argument above, we can see that it is possible to gather empirical evidence and learn from it in a *negative* way, such that if our test does not work out, we must revise our general assertion. Like the Logical Positivists, Popper was still relying on empirical evidence. But now, instead of that evidence making a difference to our experience so that it could be used for verification, evidence counted because the theory at hand was capable of being refuted by it.

Remember that black swan? If we'd seen one, it would have caused us to revise our hypothesis that "All swans are white." A single counterexample has the power—through *modus tollens*—to affect our blanket statements about the world. And that provided a way, Popper thought, for giving up on the idea of verification in science. If we wish to demarcate science from nonscience, we have to ask a simple question: is the general statement that we have made about the world capable of being refuted by *some possible experience*, even if we have not had and may never have that experience? If the answer is no, then it is not scientific.

Fortunately for Popper, a real-life example of good science was close at hand. In fact, it may have been what inspired his theory. In May 1919, Arthur Eddington set out on an expedition to take photographs of the stars during a total solar eclipse. This was crucial for the confirmation of Einstein's theory of general relativity. As Popper explains:

> Einstein's gravitational theory had led to the result that light must be attracted by heavy bodies (such as the sun), precisely as material bodies were attracted. As a consequence it could be calculated that light from a distinct fixed star whose apparent position was close to the sun would reach the earth from such a direction that the star would seem to be slightly shifted away from the sun; or, in other words, that stars close to the sun would look as if they had moved a little away from the sun, and from one another. This is a thing which cannot normally be observed since such stars are rendered invisible in the daytime by the sun's overwhelming brightness; but during an eclipse it is possible to take photographs of them. If the same constellation is photographed at night one can measure the distances on the two photographs, and check the predicted effect.... Now the impressive thing about this case is the *risk* involved in a prediction of this kind. If observation shows that the predicted effect is definitely absent, then the theory is simply refuted. The theory is *incompatible with certain possible results of observation*—in fact with results which everybody before Einstein would have expected.[10]

In other words, the falsifiability of Einstein's theory was a prime example of the proper way to do science. In one fell swoop, Popper claimed to have simultaneously solved the problem of demarcation *and* the problem of induction. That is, since science is not based on induction, it no longer mattered. He had now found a way for empirical observations to make a direct difference in testing our general assertions about the world. And, through *modus tollens*, this was deductively valid. It is important to understand that Popper was not claiming that his criterion of falsifiability was a way of demarcating meaningful from meaningless statements. Unlike the positivists, Popper did not need to use meaning as a proxy for verifiability, since he had found a direct way to tell the difference between scientific and nonscientific statements.[11] It is worth pointing out that falsifiability thus purported to identify not only what was special about science but also what was wrong with those inquiries that were merely pretending to be scientific.

We have already mentioned the example of astrology—which goes back to Popper's day—but consider here a more contemporary example. In 1981, the state of Arkansas passed Act 590, which required that public school teachers give "balanced treatment" to "creation science" and "evolution science" in the biology classroom. It is clear from the act that religious reasons were not to be offered as support for the truth of creation science, for this would violate federal law. Instead, the curriculum was expected to concentrate only on the "scientific evidence" for creation science. But was there any? And, how precisely was creation science different from creationism?

The act held that the existing situation could not stand, since teaching evolution alone could be taken as a violation of the separation between church and state, to the extent that this would be hostile to "theistic religions" and would tend to favor "Theological Liberalism, Humanism, Nontheistic religions, and Atheism in that those faiths generally include religious belief in evolution."[12] The strategy here was clear: not only were the proponents of creation science attempting to show that it was *not* religion, they were suggesting that evolution very nearly *was* religion. But since it was unacceptable to fight this battle in a religious venue, creation science advocates held that they merely wanted an equal chance to offer their views as a scientific contender to Darwin's theory of evolution by natural selection.[13]

The fate of this particular piece of legislation—and the lawsuits that followed—will be discussed later in this chapter, and revisited in chapter 8, with intelligent design theory, which took a second swing at trying to get creationism into the public schools. For now the question is a philosophical one: could falsification identify what might be wrong with creation science? Some felt that it could, for just as with the earlier claims of astrology, it seemed that the main thesis of creation science—that God created the universe and all of the species within it—was compatible with *any* evidence. Didn't the discovery of 65-million-year-old dinosaur fossils conflict with the 6,000-year timeline in the Bible? Not really, the creation scientists contended, for surely an omnipotent God could have created the entire fossil record! I hope it is clear from our earlier consideration of the problems with astrology that this sort of tendency to explain away any contrary evidence is not a shining example of falsifiability. Whereas true science goes out on a limb to test its theories against experience, creation science refused to change its theory even when there was evidence against it. Add to this the fact that creation science had precious little to offer as positive evidence in its favor, and many were willing to dismiss it as nothing more than pseudoscience.[14]

The virtues of falsification are clear. If Popper had found a way to solve the problem of demarcation, philosophers and scientists now had a powerful tool for answering the question of what is special about science. They also had a mechanism for dismissing and criticizing those practices—such as astrology and creationism—that they did *not* want to accept as scientific; if they were not falsifiable, they were not scientific. An added benefit of Popper's approach was that he had found a way for a theory to be *scientific* without necessarily having to be *true*.[15] Why did this matter? In seeking a criterion of demarcation, it mattered a great deal to those who were versed in the history of science, and understood that some of the greatest scientific minds of the last few millennia had said things that later turned out to be false. It would be wrong to think that they weren't scientists. Even though Ptolemy's geocentric theory was later replaced by Copernicus's heliocentric one, this did not mean that Ptolemy wasn't a scientist. He had based his theory on empirical data and had pushed things forward as far as he could. What mattered is that his claims were *falsifiable*, not that they were later falsified.

It would be easy to imagine that Popper's new criterion of demarcation was also a vindication of the idea of "scientific method," but that would be far from true. In fact, Popper was one of the earliest and harshest critics of the idea that there *was* such a thing as "scientific method." In his most definitive statement on the subject, appropriately titled "On the Non-Existence of Scientific Method," Popper wrote "As a rule, I begin my lectures on Scientific Method by telling my students that scientific method does not exist."[16] Elsewhere, he writes:

> The belief that science proceeds from observation to theory is still so widely and so firmly believed that my denial of it is often met with incredulity.... But in fact the belief that we can start with pure observations alone, without anything in the nature of a theory, is absurd; as may be illustrated by the story of the man who dedicated his life to natural science, wrote down everything he could observe, and bequeathed his priceless collection of observations to the Royal Society to be used as inductive evidence. This story should show us that though beetles may profitably be collected, observations may not.[17]

It is important here to remember the distinction between saying that there is a "scientific method" and saying that there is some methodological difference—such as falsifiability—between science and nonscience. Although Popper is unequivocally rejecting the idea of "scientific method," he still believes that we can have a criterion of demarcation and even one that is methodological in nature.[18]

This opinion was not shared by some of Popper's critics, notably by one of his most famous, Thomas Kuhn, who felt that although Popper was correct to abandon the idea of scientific method,[19] one should probably also give up on the idea that there is *any* distinctive methodological difference between science and nonscience. Note that this does not necessarily mean that one is giving up on the idea that science is "special" or even that there is a way of distinguishing between science and nonscience. Kuhn was not yet ready to do this (though many of his later followers were); instead, he merely pointed out that the process by which scientists actually work has much more to do with nonevidential "subjective" factors in theory choice, such as scope, simplicity, fruitfulness, and the ability to fit a theory with one's other beliefs, and much less to do with any formal method. And surely this must have an impact on justification.

It is important to understand that Kuhn was not an opponent of science. He was not—although he has been blamed for it—one of those who later claimed that science was an "irrational" process, no better than any other way of knowing, nor did he believe that the social factors that sometimes influenced scientific theory choice undermined its claim to produce credible theories. Instead, Kuhn was at pains to make sure that we understood science for what it really was, feeling that even if we did so it would be no less wonderful. While Kuhn never took it upon himself to try to provide a criterion of demarcation, he did nonetheless feel himself to be a champion of science.[20]

What of Popper's theory? Despite its virtues, it was severely criticized—by Kuhn and others—as offering too simple a picture of scientific theory change, especially given the fact that most science did not work precisely as the heroic example of Einstein's prediction indicated. There are few such crucial tests, involving risky predictions and dramatic successes, in the history of science. Most of science actually grinds along fairly slowly, with tests on a much smaller scale and, tellingly, widespread reluctance to give up a workable hypothesis just because something has gone wrong.[21] Yes, evidence counts, and one cannot simply ignore data and insulate a theory from refutation. Yet many philosophers, embracing the Duhem–Quine thesis (which says that it is always easier to sacrifice a smaller supporting hypothesis or make an ad hoc modification than to give up a theory), were skeptical that science worked as Popper said it did. Even though Popper maintained that his theory dealt only with the *logical* justification of science, many felt that there was a growing credibility gap between the way that scientists actually worked and the way that philosophers justified their work, given the sorts of social factors that Kuhn had identified. As Kuhn demonstrated, we are occasionally given to engage in a scientific revolution, but it does not happen nearly often enough for this to be accepted as the basis for demarcating between science and nonscience.

The upshot of all this is that by the 1970s, there was fairly wide agreement among most philosophers of science not only that the classic five-step scientific method was a myth, but also that there was no genuine methodological distinction between science and nonscience. This had great importance for the idea that science was special. Can one defend the idea that science is distinctive without also believing in scientific method or at least some other criterion of demarcation? Many said no.

Once Kuhn had opened the door to examining the workaday details of how scientists did their business—through "puzzle solving" and the search for accommodation to the dominant paradigm through "normal science"—the critics seemed unstoppable. To Kuhn's horror (he did after all agree with Popper and other defenders of science that *evidence counts* and that the revolution from one scientific theory to another on the basis of this evidence is the hallmark of science), his work was often cited as support by those who no longer believed that science was special. Sociologists of science, relativists, postmodernists, social constructivists, and others took their turns attacking the idea that science was rational, that it had anything to do with the pursuit of truth, or indeed that scientific theories were anything more than a reflection of the political biases about race, class, and gender held by the scientists who produced them. To some, science became an ideology, and facts and evidence were no longer automatically accepted as providing credible grounds for theory choice even in the natural sciences.

Paul Feyerabend went so far as to claim that there was no method in science at all. This was a radical departure from merely giving up on the scientific method. Along for the ride went any claims about methodology (such as objectivity), a criterion of demarcation, and even the idea that scientific beliefs were privileged.[22] Many wondered whether philosophy had now given up on science all together.

This is not to suggest that all philosophers of science felt this way. There were many who followed the ideas of Logical Empiricism (the successor to Logical Positivism), which held sway contemporaneously with Popper's theory of falsification right through the Kuhnian revolution. Here the focus was on defending the special method of science—even picking up on the earlier positivist idea of a "unified science" whose method could be extended to the social sciences—not through falsifiability (or meaning), but through close examination of how one might build more credible and reliable theories even in the face of the problem of induction.[23] Even here, though, it was necessary to modulate the full-throated defense of science, and certain concessions had to be made.[24]

By 1983, Larry Laudan, one of the most prominent philosophers of science, was ready to pull the plug on the idea that one could have a criterion of demarcation. Laudan's work was not so radical as to suggest that science wasn't important. He was one of the post-Kuhnians who looked for a way to uphold the idea that science could make "progress," though certainly not toward "true" theories or in any way that suggested the hegemony of science over other ways of knowing. In his earlier-cited article "The Demise of the Demarcation Problem," Laudan argued that there was no possible solution to the problem of demarcation, largely on the grounds that if it could be solved it would have been by now. By the time Laudan entered the picture, it goes without saying that there is no scientific method, but even the idea of finding another way of distinguishing between science and nonscience now seems dead.

Note that this does not necessarily mean that there is no difference between science and nonscience. One could even believe (as I think Laudan does) that science is uniquely explanatory. It's just that we are not going to be able to find a workable device for demarcation. Even if we all agree in our bones what is science and what is not, we are not going to be able to create a hard and fast way to distinguish it. The technical reason for this, Laudan tells us, is that philosophers have not been able to come up with a set of necessary and sufficient conditions for science. And to him this seems to be an absolute requirement for fulfilling a criterion of demarcation.

> What will the formal structure of a demarcation criterion have to look like if it is to accomplish the tasks for which it is designed? Ideally, it would specify a set of individually necessary and jointly sufficient conditions for deciding whether an activity or set of statements is scientific or unscientific. As is well known, it has not proved easy to produce a set of necessary and sufficient conditions for science. Would something less ambitious do the job? It seems unlikely. Suppose, for instance, that someone offers us a characterization which purports to be a necessary (but not sufficient) condition for scientific status. Such a condition, if acceptable, would allow us to identify certain activities as decidedly unscientific, but it would not help "fix our beliefs," because it would not specify which systems actually were scientific.... For different reasons, merely sufficient conditions are equally inadequate. If we are told, "Satisfy these conditions and you will be scientific," we are left with no machinery for determining that a certain activity or statement is nonscientific.... Without conditions which are both necessary and sufficient, we are never in a position to say 'this is scientific: but that is nonscientific.[25]

What is the problem with giving only a necessary condition? It is too strict. By aiming to exclude all that is *not* science, we may also keep out some things that we would want to include. Suppose our necessary condition were that a science must be capable of performing controlled experiments. Doesn't that rule out geology? Astronomy? All of the social sciences? Suppose, on the other hand, that we abandon this and instead aim at providing only a sufficient condition for scientific investigation: for instance, that it must be concerned with seeking truth based on empirical evidence. Here the concern is that we may have been too inclusive. Haven't we now allowed as scientific the search for Bigfoot? By trying to include everything that *is* science, we may also let in those things that we would surely want to keep out.[26] Thus, Laudan tells us, to have an adequate criterion of demarcation between science and nonscience, we need to specify a set of individually necessary and jointly sufficient conditions for science.[27]

There is perhaps no better illustration of the difficulties presented by adhering to such a lofty standard than Karl Popper's own proposal of falsifiability. Is this intended to be a necessary standard for science, a sufficient one, or both? The search for an answer is maddening. In some places, Popper writes as if he intends his account to do any one of these. Consequently, his criterion has been criticized both for excluding legitimate science (such as evolutionary biology) and for allowing some of the pseudosciences (such as astrology) to seem scientific.[28] Laudan in particular takes Popper to task for the latter, when he writes that Popper's criterion "has the untoward consequence of countenancing as 'scientific' every crank claim which makes ascertainably false assertions."[29]

This last claim would surely enrage Popper (who designed his criterion specifically to keep things like astrology *out* of the scientific pantheon). So perhaps falsification should be interpreted as providing only a necessary condition?[30] But, as we have seen, there is a weakness in this approach as well.[31] So maybe it is best to accept the idea that Popper intended to meet the highest standard by providing an "individually necessary and jointly sufficient" set of criteria. Late in life, Popper did at one point say "a sentence (or a theory) is empirical-scientific if and only if it is falsifiable."[32] In philosophy of science, those are the magic words: "if and only if" commits him to providing both necessary *and* sufficient conditions. For the reasons already given, however, falsification alone will not do, yet one looks in vain in Popper's writings for a definitive statement of what other conditions might apply. Frank Cioffi, in his important essay "Psychoanalysis, Pseudoscience, and Testability," comes close, arguing that in addition to falsifiability, Popper had intended to include the requirements that "energetic attempts are made to put the theory to test" and that "negative outcomes of the tests are accepted."[33] But even here, one runs into the previously encountered problem, identified by Kuhn and others, that negative outcomes are *not* always accepted as overthrowing a theory.

If even Karl Popper gets caught up in the problem of providing the necessary and sufficient conditions that Laudan requires for offering a criterion of demarcation, some may wonder whether the rest of us should just give up. That is precisely what occurred for almost three decades after Laudan's essay, when many were compelled by his reasoning to abandon the attempt to provide a criterion of demarcation between science and nonscience. This is not to say that they necessarily gave up on the idea that science was special. Remember that, like Laudan, one might believe that science could be defined by ostension. Surely many, like Laudan himself, were not ready to give up on the idea that science was worth defending, even if it could not be done through demarcation. With Laudan's positive account that it was still possible for science to make "progress" along Kuhnian lines, many pulled back from some of Feyerabend's and the social constructivists' extreme assertions that science was just another way of knowing. The lament instead was that even if we know science (and pseudoscience) when we see them, we are prevented from having a good way to say what defines them. Even if many had given up on the problem of demarcation, they had not given up on science.

This strategy, however, had its costs. The low point was reached in 1982, when Act 590 was challenged on constitutional grounds in the case of *McLean v. Arkansas*. The prominent philosopher of science Michael Ruse was called to testify as an expert witness and, when backed into a corner about the definition of science, gave a version of Popper's theory of falsifiability. This ended up convincing the judge, who quoted liberally from Ruse's testimony in his opinion that creationism was *not* science, and therefore had no business in the science classroom. Although Ruse had done the best he could—and was in my view unfairly criticized for being bold enough to appear in court and prevent the travesty of seeing creationism accepted as legitimate scientific theory—the academics' disapproval was swift and direct. Laudan, who surely agreed with Ruse that creationism was a farce, decried his use of Popper's theory in the judge's decision.

> In the wake of the decision in the Arkansas Creationism trial.... the friends of science are apt to be relishing the outcome.... Once the dust has settled, however, the trial in general and Judge William R. Overton's ruling in particular may come back to haunt us; for, although the verdict itself is probably to be commended, it was reached for all the wrong reasons and by a chain of argument which is hopelessly suspect. Indeed, the ruling rests on a host of misrepresentations of what science is and how it works.[34]

Laudan's worry can be summed up as follows:

> Laudan replied that creationist doctrine itself *is* science by [Popper's] criterion. It is obviously empirically testable since it has already been falsified. To be sure, its advocates have behaved in a nonscientific manner, but that is a different matter. The reason [creationism] should not be taught is simply that it is *bad* science.[35]

One imagines that if such scrupulously faithful adherence to the scholarship of the day had been allowed to hold sway in court the creationists would have been thrilled: yes, teach creationism as "bad" science, but teach it nonetheless.

Thus we see that failure to demarcate between science and nonscience can have real-world consequences outside philosophy. For one thing, the issue of teaching creationism in the public schools did not simply disappear in 1982, but instead has morphed and grown—partially as a result of philosophers' inability to defend what is special about science—into the current claim that "intelligent design (ID)" (which I have elsewhere referred to as "creationism in a cheap tuxedo") is now a full-fledged scientific theory that is ready for its debut in biology classrooms.[36] This was again put to test at trial in 2005 in *Kitzmiller v. Dover Area School District*, where another judge—in a stinging rebuke that was reminiscent of the *McLean* decision—found that intelligent design is "not science" and ordered the defendants to pay $1 million to the plaintiffs. This may give pause to future ID theorists, but sadly this story is still not over, as current "academic freedom" bills are pending in the state legislatures of Colorado, Missouri, Montana, and Oklahoma, modeled on a successful 2012 Tennessee law that defends the rights of "teachers who explore the 'scientific strengths and scientific weaknesses' of evolution and climate change."[37]

Figuring out how to demarcate between science and nonscience is no laughing matter. Being able to say, in public and in a comprehensible way, why science is special seems a particular duty for those philosophers of science who *believe in science*, but have not been able to articulate why. As climate change deniers begin to gear up, taking a page from the earlier battles of the creationists (and the tobacco lobby) in fighting scientific conclusions that they don't like through funding "junk science," then spreading it through public relations, isn't it time that we found a way to fight back?

Of late, this is precisely what has happened. In 2013, philosophers Massimo Pigliucci and Maarten Boudry published an anthology entitled *Philosophy of Pseudoscience: Reconsidering the Demarcation Problem*, in which they self-consciously seek to resurrect the problem of demarcation thirty years after Laudan's premature obituary. The papers are a treasure trove of the latest philosophical thinking on this issue, as the profession tries to steer its way out of the ditch where Laudan left it: where we believe that science is special, but can't quite say how. It is disappointing—but certainly understandable—that after all this time philosophers are a little unsure how to proceed. Perhaps resurrecting the traditional problem of demarcation is the answer. Or perhaps there is another way.

It is no small thing to dismiss the problem of demarcation, which has been the backbone of the philosophy of science since its founding. The attractiveness of using its structure and vocabulary as a way to understand and defend the distinctiveness of science is obvious. Perhaps this is why virtually all previous attempts to say what is special about science have involved trying to come up with some criterion of demarcation. But there are many pitfalls to resurrecting this approach.

In Pigliucci's essay "The Demarcation Problem: A (Belated) Response to Laudan," he rejects the "necessary and sufficient conditions" approach, preferring instead to rely on Ludwig Wittgenstein's concept of "family resemblance." Pigliucci thus claims to rescue the problem of demarcation from Laudan's "old-fashioned" approach (which may be conceived of as challenging Laudan's "meta-argument" over what is required to solve the problem of demarcation).[38] Instead Pigliucci's idea is to treat learning the difference between science and pseudoscience as a kind of "language game," where we come to learn the clusters of similarity and difference between different concepts by seeing how they are used. The goal here is to identify the various threads of relationship that do not fall neatly along the lines of necessary and sufficient conditions but nonetheless characterize what we mean when we say that some particular inquiry is scientific. Two of these threads—"empirical knowledge" and "theoretical understanding"—appear to do most of the work. As Pigliucci writes, "if there is anything we can all agree on about science, it is that science attempts to give an empirically based theoretical understanding of the world, so that a scientific theory has to have both empirical support and internal coherence and logic."[39] As a result, Pigliucci thinks that—among other things—we will have discovered a "Wittgensteinian family resemblance" for the concepts of science and pseudoscience that provides a viable demarcation criterion to "recover much (though not necessarily all) of the intuitive classification of sciences and pseudosciences generally accepted by practicing scientists and many philosophers of science."[40]

This account, however, seems quite nebulous as a criterion of demarcation. For one thing, what is its logical basis? At various points Pigliucci refers to the use of "fuzzy logic" (which relies on inferring degrees of membership for inclusion in a set) to help make his criterion more rigorous, but it remains unclear how this would work. As Pigliucci admits, "for this to actually work, one would have to develop quantitative metrics of the relevant variables. While such development is certainly possible, the details would hardly be uncontroversial."[41] To say the least: one imagines that the central concepts of empirical knowledge and theoretical understanding may be equally as difficult to describe and differentiate from their opposites as the concept of science itself. Has Pigliucci solved the problem of demarcation or merely pushed it back a step?

Others who have pursued a "post-Laudan" solution to the problem of demarcation have encountered a similarly rocky path. In the same volume, Sven Hansson pursues an extremely broad definition of science. Apparently wary of the implications of classifying disciplines like philosophy as nonscience, he instead expands the scope of science to mean something more like a "community of knowledge" and then proceeds to demarcate this from pseudoscience. For all of the alleged advantages of rescuing the humanities from the realm of nonscience, however, the cost is quite high, for now he cannot say that the problem with pseudoscience has anything to do with its bastardization of empirical standards (since at least part of what he now classifies as science is not empirical either).[42]

Maarten Boudry takes a similarly questionable step in saying that he thinks there are really two demarcation problems—the "territorial" and the "normative"—instead of just one. The former dispute he dismisses as sterile. It is just a matter of "turf" that concerns separating science from legitimate but nonempirical epistemic endeavors like history and philosophy. According to Boudry, the real dispute is between science and pseudoscience; this is where the normative issue arises, because this is where we face those disciplines that are just pretending to be sciences.[43] Yet this bifurcation of the problem of demarcation reveals a basic confusion between saying that a discipline is *non*scientific and saying that it is *un*scientific. Does Boudry mean to identify the "territorial" dispute as one between fields that are scientific and *non*scientific? If so, that is a highly idiosyncratic and misleading use of the term. The dispute that he appears to be searching for when he talks about the territorial problem of demarcation seems to be between science and what may be called "*un*science." Yet why would this be the proper alternative to the normative dispute? The more traditional interpretation of the demarcation debate—revealed in most scholarship in the field—is that between science and *non*science, or between science and *pseudo*science. These are the terms of art used by Popper, Laudan, and most everyone else.[44] Instead, Boudry seems to be creating a *new* demarcation problem, while saying nothing about why we should ignore the classic problem of demarcating between science and *non*science. But why does Boudry think that he can make a case for the normative battle between science and *pseudo*science, when he has not legitimately dispensed with the larger issue of science versus *non*science? The straw-man "territorial" distinction between science versus *un*science (history, philosophy, etc.) does not do the job.[45]

This struggle to explain whether the problem of demarcation should be between science versus *non*science—or between science versus *pseudo*science—may seem like a mere terminological dispute, but it is not. For if we are attempting to distinguish science from *all that is not science*, it may lead to a very different criterion of demarcation than if we seek to distinguish science *merely from its imposter*. The important point here is to recognize that, according to most scholars, the category of nonscience *includes* both those fields that are pseudo-scientific and those that are unscientific. An inquiry can be nonscientific either because it is merely pretending to be scientific (in which case it is *pseudo*scientific) or because it concerns matters where empirical data are not relevant (in which case it is *un*scientific).[46] (See figure 1.1.)

This failure to be specific about what one is differentiating science from, however, not only exists in the "post-Laudan" essays by Pigliucci, Hansson, and Boudry, but seems to reflect a deep equivocation in the literature that goes all the way back through Laudan to Karl Popper himself. Remember that in *The Logic of Scientific Discovery*, Popper says that he is demarcating science from math, logic, and "metaphysical speculation."[47] By the time he gets to *Conjectures and Refutations*, however, his target is pseudoscience. In Laudan's essay, he too slides back and forth between talking about nonscience and pseudoscience.[48]

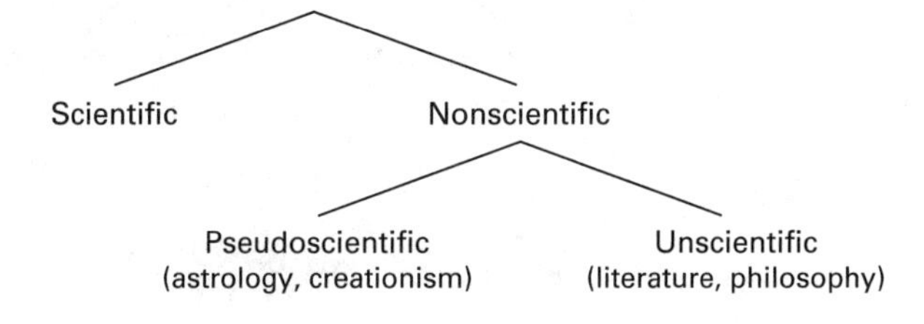

Figure 1.1

What difference does all of this make? It turns out to be crucial. Later we will revisit the issue of necessary and sufficient conditions and learn that the entire question of what is special about science may hang in the balance. We will see that the project of trying to solve the problem of demarcation is hamstrung unless we can specify precisely what it is that we are trying to define (science, nonscience, pseudoscience, or unscience) and, as we have seen, there has as yet been no definitive answer to that. My goal will be to provide a way to say what is distinctive about science without getting tripped up on the problem of providing both necessary and sufficient conditions—or trying to solve the problem of demarcation—because I do not think that these problems can be solved. Yet we still need a way to defend science.

First, however, let us deal with the problem of those who have misunderstood how science works.

2 Misconceptions about How Science Works

It is a popular myth that science inevitably leads to truth because it uses empirical evidence to prove a theory. It is another that science has no bearing at all on what we should believe because everything it proposes is "just a theory." Actually, these two misconceptions go hand in hand, for they seem to reflect the idea that science is all or nothing: that we can either be 100 percent certain that our theory has been verified by the evidence or we are completely at sea because—until the definitive experiment has been done—one theory is just as good as any other.

This false idea that science must discover truth, and that a theory cannot be accepted into the scientific pantheon until it is absolutely verified by the data, means that those who do not understand science might feel justified in rejecting any scientific knowledge that falls short of this standard. But this is a radical misunderstanding of how science works.

The Problem with Truth and Certainty

As with most misconceptions, there is a kernel of truth in the critics' case that has gone awry. Science *aims* at truth. It does this by attempting to test its theories rigorously against empirical data. The vetting is savage. As we saw with Karl Popper, if a theory meets with falsifying data, there is a problem. We may not know at the outset whether the theory can be saved through judicious modifications but, on pain of ceasing to be scientific, eventually something must be done. Yet even if a theory passes every test with flying colors, we *still* cannot be certain that it is true. Why not? Because, as we shall see in this chapter, that is not how science *actually* works. The only thing we can be certain of in science is that when a theory fails to be consistent with the empirical evidence there must be something wrong, either with the theory itself or with one of the auxiliary assumptions that support it. But even when a theory *is* consistent with the evidence, we can never be sure that this is because the theory is true or merely because it has worked out so far.

As Popper, Kuhn, and many others in the philosophy of science long ago recognized, scientific theories are always tentative. And that is the foundation for both the strength and the flexibility of scientific reasoning. Any time we are dealing with empirical data, we face the problem that our knowledge is open ended because it is always subject to revision based on future experience. The problem of induction (which we met briefly in the last chapter) is this: if we are trying to formulate a hypothesis about how the world works, and are basing this hypothesis on the data that we have examined so far, we are making a rather large assumption that future data will conform to what we have experienced in the past. But how do we know this? Just because all of the swans we have seen in the past are white, this does not preclude the existence of a black swan in the future. The problem here is a deep one, for it undermines not just the idea that we can be *certain* that any of our proposals about the world are true (no matter how well they may conform to the data), but also, technically speaking, we cannot even be sure that our proposals are more *likely* to be true, given the indefinably small relationship between the size of the sample of the world we have examined so far compared to the size of the set of possible experiences we may have in the future. How can we be sure that the sample of the world we have seen so far is representative of the rest of it? Just as we cannot be sure that the future is going to be like the past, we cannot be sure that the piece of the world we have met in our limited experience can tell us anything at all about what it is like elsewhere.

Naturally, there is a remaining debate here (courtesy of Karl Popper) over whether science actually uses induction. Even though science tries to draw general conclusions about how the world works based on our knowledge of particular circumstances, Popper proposed a way to do this that avoids the uncertainties of the problem of induction. As we saw in chapter 1, if we use *modus tollens*, then we can learn from data in a way that is deductively valid. If rather than trying to verify our theory we instead try to falsify it, we will be on more solid logical footing.

But it is now time to realize that this way of proceeding faces deep criticisms. For all of the logical virtues of Karl Popper's theory, it is questionable (1) whether his account does justice to the way that science is actually carried out and (2) whether he can avoid becoming entangled in the need to rely on positive (confirming) instances. The Duhem–Quine thesis says that it is impossible to perform a "crucial test" in science, because every theory exists within a web of supporting assumptions. This means that even when there are falsifying instances for a theory, we can always sacrifice one of its auxiliary assumptions in order to save it. According to strict falsificationism, this must seem wrong. Popper himself, however, claims that he had already anticipated this objection and accommodated it within his theory.[1] According to the logic of falsification, when a theory is falsified we must abandon it. But in actual practice, Popper recognized, scientists are loath to give up a well-ensconced theory just because of one falsifying instance. Maybe they made a mistake; maybe there was something wrong with their apparatus. Popper admits as much when he writes "he who gives up his theory too easily in the face of apparent refutations will never discover the possibilities inherent in his theory."[2]

In explicating his falsificationist account, Popper nonetheless prefers to illustrate its virtues with stories of those larger-than-life dramatic instances when a theorist made a risky prediction that was later vindicated by the data. As we've seen, Einstein's general theory of relativity made a bold prediction about the bending of light in a strong gravitational field, which was confirmed during the total solar eclipse of 1919. If the prediction had been wrong, the theory would have been rejected. But, since it was right, the epistemic reward was tremendous.[3]

But most science does not work like this. In his book *Philosophy of Science: A Very Short Introduction*, Samir Okasha recounts the example of John Couch Adams and Urbain Le Verrier, who discovered the planet Neptune in 1846, when they were working (independently) within the Newtonian paradigm and noticed a slight perturbation in the orbit of the planet Uranus. Without Neptune, this would have been a falsifying instance for Newtonian theory, which held (following Kepler) that all planets should move in perfectly elliptical orbits, unless some other force was acting upon them. But, instead of abandoning Newton, Adams and Le Verrier instead sought out and found that other gravitational force.[4]

Some might complain that this is not even close to a falsifying instance, since the theorists were working well within the predictions that had been made by Newtonian theory. Indeed, Popper himself has sometimes cited this very example as just such an instance where scientists were wise not to give up on a theory too quickly. Yet contrast this with a similar challenge to Newtonian theory that had been around for over one hundred and fifty years by the time of the discovery of Neptune: the slight perihelion advance in the orbit of the planet Mercury.[5] How could this be explained? Astronomers tried various ad hoc solutions (sacrificing auxiliary assumptions) but none were successful. Finally, it was none other than Le Verrier himself who proposed that the slight perturbation in Mercury's orbit could be explained by the existence of an unseen planet—which he named Vulcan—between Mercury and the Sun. Although his tests to find it were unsuccessful, Le Verrier went to his grave in 1877 believing that Vulcan existed. Virtually all of the other astronomers at the time agreed with him in principle that—whether Vulcan existed or not—there must be some Newtonian explanation. Forty years later Einstein pulled this thread a little harder and the whole Newtonian sleeve fell off, for this was an instance in which the anomalous orbit was explained not by the strong gravitational force of another *planet*, but instead by the non-Newtonian idea that the *Sun*'s gravitational force could warp the space around it. When the orbit of Mercury fit those calculations, it ended up being a major mark in favor of the general theory of relativity. Yet this was not a prediction but a "retrodiction," which is to say that Einstein's theory was used to explain a past falsifying instance that Newtonian theorists *had been living with for over two hundred years*! How long is too long before one "gives up too easily" in the face of a falsified theory? Popper provides no rules for judging this. Falsification may work well with the logic of science, but it gives few guidelines for how scientists should actually choose between theories.

As Thomas Kuhn demonstrated, there are no easy answers to the question of when a falsifying instance should take down a well-believed theory and when it should merely lead us to keep searching for answers within the dominant paradigm. In Kuhn's work, we find a rich description of the way that scientists actually struggle with the day-to-day puzzles that present themselves in the course of "normal science," which is when we work to accommodate the predictions, errors, and what may seem like falsifying instances within the four corners of whatever well-accepted theory one may be working under at the time.[6] Of course, Kuhn also recognized that sometimes science does take a turn toward the dramatic. When anomalies pile up and scientists begin to have a hard time reconciling their paradigm with things it cannot explain, enough force builds up that a scientific revolution may occur, as the field shifts rapidly from one paradigm to another. But, as Kuhn tells us, this is often about more than mere lack of fit with the evidence; it can also encompass scope, simplicity, fruitfulness, and other "subjective" or "social" factors that Popper would be reluctant to account for within his logical account of falsification.

But there are other problems for Popper's theory, too, even if a hypothesis *passes* all of the rigorous tests we can throw at it. As Popper admits, even when a theory succeeds it cannot be accepted as true—or even approximately true—but must always inhabit the purgatory of having merely survived "so far."[7] As powerful as scientific testing is, we are always left at the end with a potentially infinite number of hypotheses that *could* fit the data, and an infinite supply of possible empirical evidence that *could* overthrow any theory. Scientific reasoning thus is forced to make peace with the fact that it will always be open ended, because the data will always be open ended too. Even if we are not "inductivists" about our method, we must admit that even when a theory has passed many tests there are always going to be more ahead.

Popper tried to solve this problem through his account of corroboration, in which he said that after a theory had survived many rigorous tests, it built up a kind of credibility such that we would be foolish to abandon it cavalierly; as Popper put it, some theories have "proven their mettle."[8] To some ears, however, this began to sound very much like the type of verification and confirmation that Popper insisted he had abandoned. Of course, we cannot say that a theory is *true* just because it has passed many tests. Popper admits as much. But the problem is that we also cannot say that a theory is even more *likely* to be true on the basis of these tests either. At some level, Popper seems to understand this danger (as well he should, for this is just the problem of induction again), but it is unclear how he proposes to deal with it.[9] Remember that induction undermines not just certainty, but also probability. If the sample of potential tests is infinite, then the sample we've chosen to test our theory against is infinitesimally small. Thus it cannot help a theory's verisimilitude that it is well corroborated. At various points, Popper says that falsification is a "purely logical affair."[10] But why then does he introduce a notion like corroboration? Perhaps this is a rare concession to the practical issues faced by scientists? Yet what to make then of Popper's claim that falsification is not concerned with practical matters?[11]

Philosophers of science will continue to fight over the soul of Karl Popper. Meanwhile, many are left to draw the inevitable conclusion that inductivist-type problems cannot help but accrete for surviving theories, even if one is a falsificationist. The idea that there are always more data that can overthrow a theory—along with the related idea that there are infinitely many potential theories to explain the evidence that we already have—is frequently dismissed as a philosopher's worry by many scientists, who continue to maintain that when a theory survives rigorous testing it is either true or more likely to be true. But in their bones I am not sure how they could possibly believe this, for scientists as well as philosophers of science know full well that the history of science is littered with the wreckage of such hubris.[12] Perhaps this is one instance in which some of those who wish to defend science may be tempted to pretend that science *can* prove a theory, even if they know it cannot. Sometimes, in the excitement of discovery, or the heat of criticism, it may seem expedient to maintain that one's theory is *true*—that certainty is possible. But I submit that those who wish to defend science have a special obligation to accept the peculiarities of science for what they are and not retreat behind half-truths and wishful thinking when called upon to articulate what is most special about it. Whether we believe that science is based on inductive reasoning or not—whether we believe in falsificationism or not—we must accept that (no matter how good the evidence) science *cannot* prove the truth of any empirical theory, nor can we say that it is even, technically speaking, more probably true.

Yet as we shall see, this does not mean that we have no grounds for believing a scientific theory.

"Just a Theory"

At this point it is important to confront a second misconception about science, which is that if a scientific claim falls short of "proof," "truth," or "verification" then it is "just a theory" and should not be believed. Sometimes this is read out as a claim that another theory is "just as likely to be true" or "could be true," while others maintain that any theoretical knowledge is simply inferior.

The first thing to understand is that there is a difference between a theory and a hypothesis. A hypothesis is in some ways a guess. It is not normally a wild guess; it is usually informed by some prior experience with the subject matter in question. Normally a hypothesis arises after someone has noticed the beginnings of a pattern in the data and said, "Huh, that's funny, I wonder if..." Then comes the prediction and the test. A hypothesis has perhaps been "back-tested" by our reason to see if it fits with the data we have encountered so far, but this is a far cry from the sort of scrutiny that will meet a hypothesis once it has become a theory.

A scientific theory must not only be firmly embedded in empirical evidence, it must also be capable of predictions that can be extrapolated into the wider world, so that we can see whether it survives a rigorous comparison with *new* evidence. The standards are high. Many theories meet their demise before they are even put forth by their proponents, as a result of meticulous self-testing in advance of peer review. Customarily a theory must also include an explanatory account for *why* the predictions are expected to work, so that there is a way to reason back from any falsification of a prediction to what might be wrong with the theory that produced it. (As we saw with the example of the perihelion advance of Mercury, it is also a plus if a theory can explain or retrodict any anomalies that scientists have been contending with in their previous theory.)

Here we should return to Karl Popper for a moment and give him some credit. Although he may not have been right in the details of his account of falsifiability—or his more general idea that one can demarcate science from nonscience on a logical basis—he did at least capture one essential element that explains how science works: our knowledge of the world grows by keeping close to the empirical evidence. We can love a theory only provisionally and must be willing to abandon it either when it is refuted or when the data favor a better one. This is to say that one of the most special things about science is that *evidence counts*.

The Nobel Prize–winning physicist Richard Feynman said it best:

> In general, we look for a new law by the following process: First we guess it. ... Then we compute the consequences of the guess to see what ... it would imply. And then we compare the computation results to nature, or we say compare to experiment or experience, compare it directly with observations to see if it works. If it disagrees with experiment, it's wrong. In that simple statement is the key to science. It doesn't make any difference how beautiful your guess is, it doesn't make any difference how smart you are, who made the guess, or what his name is. If it disagrees with experiment, it's wrong. That's all there is to it.[13]

In this sense, it is not entirely wrong to say that "scientific method" *does* capture something important about the process of scientific reasoning. Even if it does not quite serve as a criterion of demarcation, it does demonstrate the *critical state of mind* that one must have when testing a theory against sensory evidence—that is the hallmark of empirical knowledge. We see something strange, we make a hypothesis about it, we make a prediction, we test it, and if all goes well we may have a theory on our hands.[14] This kind of reasoning may not be *unique* to science, but without it one can hardly see how science could go forward.

A theory arises when we are prepared to extrapolate a hypothesis more widely into the world. A theory is bigger than a hypothesis because it is the result of a hypothesis that has been shaped by its clash with the data, and has survived rigorous testing before someone puts it forward. In some sense, a theory is just short of a law of nature. And you can't get any more solid than a law of nature. Indeed, some have held that this is what scientists have been searching for all along when they say that they are searching for the "truth" about the empirical world. They want to discover a scientific law that unifies, predicts, and explains the world we observe. But laws have to be embedded in a theory. And a theory has to be more than a guess. A theory is the result of enormous "beta testing" of a hypothesis against the data, and projection of a reason *why* the pattern should hold for future experience. The apple falls *because* it is pulled by gravity. The global temperature is rising *because* of greenhouse gas emission. A scientific theory seeks to explain both how and why we see what we see, and why we will see such things in the future. A theory offers not just a prediction but an explanation that is embedded in the fabric of our experience.

Ideally, a scientific theory should (1) identify a pattern in our experience, (2) support predictions of that pattern into the future, and (3) explain why that pattern exists. In this way, a theory is the backbone of the entire edifice of scientific explanation. For example, when one looks at Newton's theory of gravity, it is remarkable that it unified both Galileo's theory of terrestrial motion and Kepler's theory of celestial motion. No longer did we have to wonder why objects fell to Earth or planets orbited the Sun; both were accounted for by the law of gravity. This fit well with our experience and explained why, if you throw a ball near the surface of the Earth, it will—like the planets—trace an elliptical path (and that if you threw it hard enough it would go into orbit), and supported predictions as well (for instance, the appearance and return of comets). Did it give an account of how gravity did this? Not quite yet. Newton famously said that he "framed no hypotheses" on the subject of what gravity was, leaving it for others later to deal with the anomalies of action at a distance and how attraction and repulsion could occur through empty space. He had a theory, but not yet a mechanism.

Doesn't this undermine the idea that a theory explains because it tells us *why* things happen? This is controversial. Some of the most famous scientific explanations could not say at the time they were offered what causal powers were responsible for the patterns they explained. (Another outstanding example here is Darwin's theory of evolution by natural selection, which awaited Mendelian genetics before it could say why evolution occurred.)[15]

This raises the question of whether scientific theories are just instruments for prediction. Whether they are a mere shorthand account of patterns in our experience that—given the limits of scientific theorizing—can never offer a definitive answer to the mechanism behind them. It is generally thought that this is not enough for scientific explanation—that scientific theories must try to say not just *that* something happened but *why*. The answers don't have to be immediate, but a good theory should give some promise that answers will be available upon further empirical investigation. The importance of this is illustrated by what happens when we do *not* really have a theory at stake: when all we have are beautiful predictions, but no explanation for why the predictions are fulfilled.

"Bode's law" is one of the most dramatic illustrations in the history of science of just how far you can push things (and how quickly they can come crashing down) if you have good fit with the evidence—and even a few good predictions—but no theoretical support. In 1772, after studying the distance between the planets for what must have been a very long time, Johann Bode noticed a startling correlation. If one takes the common doubling series {0, 3, 6, 12, 24, 48, 96, 192, 384, 768}, adds 4 to each number, and then divides by 10, one gets {0.4, 0.7, 1.0, 1.6, 2.8, 5.2, 10.0, 19.6, 38.8, 77.2}, which is almost identical to the distance of the planets from the Sun measured in astronomical units (an astronomical unit is defined as the distance between the Earth and the Sun). In 1772, there were only six known planets, which had the following distance from the Sun: Mercury (0.387), Venus (0.723), Earth (1.0), Mars (1.524), Jupiter (5.203), and Saturn (9.539). At first, no one seemed bothered by the lack of an explanatory mechanism;

hadn't Newton after all famously "framed no hypotheses" about gravity? Still, there were questions. What about the "gap" at 2.8? And what about the rest of the series? These were taken to be "predictions" that would indicate where further planets might be. When the planet Uranus was discovered nine years later at 19.18 astronomical units, people were agog. Twenty years later, when scientists came to believe that there had once been a planet (which they posthumously named "Ceres") between Mars and Jupiter, that had broken up and formed the asteroid belt at 2.77 astronomical units, Bode's law was hailed as a momentous scientific achievement. Although it *explained nothing* (because it had virtually no scientific theory for its predictions to confirm), it was taken seriously because it had successfully predicted two new planets. When Neptune was discovered at 30.6 astronomical units in 1846, followed by Pluto at 39.4 astronomical units in 1930, things began to fall apart. Bode's law was finally accepted as nothing more than a remarkable artifact of naive correlation.[16]

Compare this with something like string theory, which in its most modest rendering is a theory of gravity at the microscopic level and at its most ambitious is a complete theory of everything in the universe.[17] Entire books are devoted to the technical details of this fiendishly difficult subject, but the short story is this. Einstein's general theory of relativity proposes to explain the largest things in the universe (stars and galaxies), while quantum mechanics proposes to explain the smallest (molecules and atoms).[18] Both theories are incredibly well-supported by the empirical evidence, but there is a problem: they are fundamentally incompatible with one another. To put it bluntly, they cannot both be right. It may, however, be the case that while neither one is completely correct, both theories are special cases of some larger theory that subsumes and explains the phenomena covered by each of them. One candidate for such a theory—known as the Standard Model in physics—has done a good job of accounting for all of the fundamental forces in the universe but one: gravity. This has led physicists on an ambitious search for a quantum theory of gravity, of which string theory has been held to be the most promising (but not the only) candidate. But there is another problem: string theory has absolutely no empirical support to suggest that it is right.

At this point, string theory is a mathematical model that many physicists hope is correct, for the simple reason that there are few alternatives. But this raises an important question: without empirical support, is string theory even science? Isn't it "just a theory"?[19] We here face a situation that is the direct opposite of the one just confronted with Bode's law: instead of an explanation that has amazing fit with the data, but no theory behind it, we instead have an incredibly complex and fruitful theory with absolutely no empirical support. But doesn't this violate our earlier criterion that a scientific theory must be vetted against some sort of evidence?

Precisely this question was considered at an academic conference entitled "Why Trust a Theory? Reconsidering Scientific Methodology in Light of Modern Physics," held at the Ludwig Maximilian University in Munich, Germany, in December 2015. This unusual conference brought together both physicists and philosophers to consider whether it was possible for there to be a new way of doing science. One such proposal was made by the physicist-turned-philosopher Richard Dawid, whose book *String Theory and the Scientific Method* was sanguine about the idea that, given the difficulty of gathering empirical support for string theory, we needed to turn to alternative modes of "non-empirical theory assessment" such as explanatory coherence, unification, fruitfulness, and even the aesthetic criteria of "elegance" or "beauty."[20] This was shot down by several of the scientists in attendance, who argued that even though string theory may not have any *currently testable* empirical consequences (because of the enormous practical limitations on building the apparatus to test them), it does make predictions that are *in principle* testable.[21]

Not all physicists would agree with this. Some have even held that there are sociological factors at work such as "groupthink" and the pressures of tenure, career advancement, and grant money at stake, which—even absent empirical support—have made string theory "the only game in town."[22] To say such things, however, seems different from saying that string theory is not *testable*. If one reads these kinds of criticisms closely one finds careful phrasing that string theory "makes no predictions about physical phenomena at experimentally accessible energies" and that "at the moment string theory cannot be falsified by any conceivable experimental result."[23] But these are weasel words, born of scientists who are not used to taking seriously the distinction between saying that a theory is "currently" testable versus whether it is "in principle" testable. The practical limitations may be all but insurmountable, but philosophical distinctions like demarcation live in that difference.

Perhaps the critics are right and string theory gets far too much attention given its current lack of empirical support. Almost certainly they are right that it would be absurd to "redefine science" in order to accommodate this one theory. Time will tell whether the paradigm of string theory has enough going in its favor to survive—as a practical matter—without the buttress of empirical support. But, on the question of whether it is science, I come down on the side of those who draw a distinction between saying that something is not testable "now" versus saying that it is not testable "in principle" (for instance, if it made no empirical predictions). Which means that—like so many other theories before it—string theory may be scientific, even if it later turns out to be wrong.

What does all of this mean for the more general issue of the importance of theories for science? Having a theory may be essential for science, but is it enough? Doesn't there have to be some evidence in support of a theory—or at least some possible evidence that *could* support it—before we can claim that it is scientific? If not, how will we ever be able to say why nature works as it does, which seems necessary for scientific explanation? As we saw with Ptolemy and Newton, a theory can be scientific and false, but it must seek to fit with the evidence and make some attempt to explain it. We do not need to take the drastic step of rejecting every false theory as unscientific. What makes a theory scientific isn't that it is true, it is that it says something—even if it is just a promissory note to be cashed in the future—about whether there is a mechanism behind the theory that supports its predictions and is consonant with empirical evidence down the line. Are we there yet with string theory? Will we ever be? Scholars disagree. But we may nonetheless appreciate that without a theory, we would not even be having this conversation. This is why Bode's law failed and string theory may yet succeed.

Why then do some critics complain that everything in science is "just a theory"? Is it because they believe that every scientific theory is as controversial as string theory? Or is it that they just don't understand (or don't *want* to understand) how powerful it is to say that we have a good scientific theory?

The theory of evolution by natural selection is "just a theory," but it is instantiated in virtually everything we believe in microbiology and molecular biology, from the cellular level up to species. It has been rigorously tested for a hundred and fifty years. It accounts for the data, makes predictions, and is completely unified with Mendelian genetics, which is the mechanism behind it. Evolutionary theory is the absolute backbone of scientific explanation in biology. Indeed, the famous evolutionary biologist Theodosius Dobzhansky was taken by many to speak for the profession when he said: "Nothing in biology makes sense except in the light of evolution."[24]

The alleged holes in Darwin's theory of evolution by natural selection are in some cases nothing more than misunderstandings by the layperson of what biology is about. And any actual holes are nothing more than the sorts of research problems that one would expect scientists to be working on in such a mature science. As Kuhn taught us, in any open-ended enterprise you will never have explained everything. One must keep pushing forward.[25] But make no mistake. Scientists have accounted for the complexity of the eye. They have even found a candidate for the "missing link."[26] The sort of nonsense one sees put forth by creationists in an attempt to discredit evolution are not the sorts of criticisms that scientists make to one another; they are the stuff of ideologues and conspiracy theorists.[27]

Gravity too is "just a theory." So is the germ theory of disease. And the heliocentric solar system. Indeed, as we have seen, everything in science is "just a theory." But this does not mean that we have no reason for belief. Having a good theory is the foundation for science. We don't need deductive certainty for a theory to be scientific or for it to be believable. The notion here is a delicate one but nonetheless important: we are entitled to believe a theory based on the evidence that supports it, while knowing full well that any future evidence may force us to give up our beliefs and change to another theory. In science we must simply do the best we can with a rigorous analysis of the data that we have available. Our beliefs can be justified, even if we cannot (and should not) quite bring ourselves to maintain that they are true.

Thus in some sense the critics are right. Science *cannot* prove anything. And everything that science proposes *is* just a theory. When we are at the mercy of future data, this is the situation that *all* empirical reasoning must contend with. And, unfortunately, this is the basis on which some members of the public—most particularly the ideological critics of science—have misunderstood how science works. It is true that science must contend with the open-ended nature of empirical reasoning, yet it is also rigorous, meticulous, and our best hope for gaining knowledge about the empirical world. But how can this be?

The Role of Warrant

It is now time to introduce the concept of *warrant* into the debate about whether we are justified in believing empirical theories. Even though they are not provably true, certain, or (even in principle) more likely to be true, there is a sense in which we would be foolish to ignore the idea that scientific theories are believable precisely because they have positive empirical support.

Here a crucial distinction must be drawn between *truth* and *warrant*. Even if a scientific theory is not technically, logically, more likely to be true once it has survived a number of rigorous tests, the question arises "aren't we nonetheless justified in *believing* it?" And here I think one may plausibly say that the answer is yes. Despite the logical problems presented by induction, verification, and confirmation (and seeing how easily Popper's practical concession of corroboration can slip into relying on positive instances), there is something deeply important about the success of empirical tests that scientists seem justified in wanting to hang on to. The credibility of a scientific theory *does* seem to increase once it has survived a number of rigorous tests. Indeed, even most philosophers of science—who are very familiar with all of the problems of inductive logic—understand that it would be rash to claim that just because positive evidence cannot be used to prove that a theory is *true*, this means that the theory is not *believable*.[28]

There is a subtle distinction between saying that a theory is true and saying that we are justified in believing it. The idea goes back to Socrates. Perhaps we may never reach the truth in our lifetimes, but can we make progress toward it? Can we at least eliminate false claims to knowledge? To say that a theory has warrant is to say that it has a credible claim to believability; that it is justifiable *given the evidence*. This is to say that even if a theory later turns out to be mistaken—like Newton's theory of gravitation—one may still maintain that *given the evidence at the time*, scientists were rational to believe it. Why is this important? Precisely because—given the way that science works—one expects that in the long run *virtually all of our empirical theories will turn out to be false*.[29] But this does not mean that we are unscientific for believing them, or that it would be better to withhold all belief until the "rest of the evidence" is in. Indeed, given how science operates, the rest of the evidence will *never* be in!

The doctrine of *fallibilism* accepts that we can never be certain of any empirical belief, yet maintains that it is unreasonable to think that all knowledge requires certainty.[30] Yes, the problem of induction undermines both certainty and probability, but what sort of epistemological stance is appropriate in the face of this? Should we give up all nondeductive beliefs? Even where the empirical evidence is strong, should we refuse to say that we *know* anything? This seems absurd. The fallibilist accepts that—outside deductive logic and mathematics—we can never achieve certainty. But this does not mean that we must foreswear all claims to knowledge. Not everything that is true is necessarily true.[31] And at least some of these nonnecessary truths are surely worth pursuing. Instead of giving up on vast tracts of possible knowledge, perhaps we should enlarge the notion of knowing so that it includes the idea that we can have justified empirical beliefs, even if we also understand that some of these beliefs may later turn out to be false. The doctrine of fallibilism, therefore, is as much an attitude as a set of principles: it tells us that it is all right to feel comfortable with the idea that some of our beliefs are justified based on their fit with the evidence, even if in the long run they may turn out to be mistaken.

Of course, it is important not to be gullible or overambitious. We must avoid thinking that just because we have good current evidence for an empirical belief, it is probably true; that although we may not *know* that we know something, the strength of the evidence suggests that the underlying theory must be pretty close to reality.[32] Yet, in compensation for this epistemological equanimity, we are saved from having to retreat into a sterile skepticism, where we take the problem of induction so seriously that we can believe in nothing, because it might someday be overthrown by better evidence. The idea of warrant can be squared with reliance on empirical evidence, even if fallibilism also requires that no accumulation of evidence will ever amount to certainty.

We cannot hold reasoned belief hostage to certainty. Despite Popper's and others' best efforts, the process of scientific reasoning is just never going to be deductively valid. Scientists are right to want to rely on the encouragement they get from positive instances, as long as they do not go overboard and assert that their theory is true, or indulge in the temptation to overlook negative evidence. But what about the problem of induction? I may not be the first philosopher to think this, but I will take the risk of saying here what I have heard many philosophers say in private: to hell with the problem of induction. The problem of induction was never meant to substitute mental paralysis for justified belief. Even David Hume seemed to recognize that in some sense induction is wired into how humans reason:

> Most fortunately it happens, that since reason is incapable of dispelling these clouds, nature herself suffices to that purpose, and cures me of this philosophical melancholy and delirium, either by relaxing this bent of mind, or by some avocation, and lively impression of my senses, which obliterate all these chimeras.

> I dine, I play a game of backgammon, I converse, and am merry with my friends; and when after three or four hours' amusement, I would return to these speculations, they appear so cold, and strained, and ridiculous, that I cannot find in my heart to enter into them any farther.[33]

Both our brains and our instinct tell us that positive instances count. As human beings, we cannot help ourselves in relying on induction to learn about the empirical world.

Yet might this be appropriate after all? One of the most fascinating responses to the problem of induction was given by Hans Reichenbach, who made the case that even if induction could not be logically verified, it could at least be vindicated.[34] His argument is intriguing. Suppose that the world is disordered: that there are no empirical correlations between anything. In this situation, no method would be able to account for or explain the regularities in our experience, because there would be none. Consider, on the other hand, a world more like our own, in which there were correlations in our experience and we wished to employ a method that would pick up on them. What method should we use? Arguably, the best choice would be induction. While it is possible that other methods might work, none would work better than induction. Like science itself, induction is responsive to patterns in our experience, flexible enough to change its conclusions based on new evidence, and capable of abandoning any individual hypothesis and starting again if the data warrant it. Even though induction may sometimes lead us into error, *so would any other method of reasoning*. And, although some other method may work, *it will never surpass induction*. Reichenbach thus concludes that induction will do at least as well as any other method in discovering the regularities in the empirical world. Thus inductive reasoning is "pragmatically vindicated."[35]

Might a similar move now be made in defense of the notion of warrant? As we have seen, what gives a scientific theory warrant is not the certainty that it is true, but the fact that it has empirical evidence in its favor that makes it a highly justified choice in light of the evidence. Call this the *pragmatic vindication* of warranted belief: *a scientific theory is warranted if and only if it is at least as well supported by the evidence as any of its empirically equivalent alternatives*. If another theory is better, then believe that one. But if not, then it is reasonable to continue to believe in our current theory.[36] Warrant comes in degrees; it is not all or nothing. It is rational to believe in a theory that falls short of certainty, as long as it is at least as good or better than its

3 The Importance of the Scientific Attitude

Many thinkers have tried to identify what is special about science based on its allegedly unique methodology. This approach has been criticized because it has been shown that many scientists do not actually follow the steps that philosophers of science have used to justify their work.[1] This does not mean that there is nothing important about what scientists do that might have great bearing on science's privileged epistemic status. It is just that perhaps we should be looking less at the method by which science is justified and more at the attitude that its practitioners have in mind while they are doing it.

As we saw in chapter 1, there is no recipe for doing science. Likewise, there may be no logical means of distinguishing between the type of reasoning that scientists use to think about the empirical world and that which is used elsewhere. Nonscientists can certainly be rigorous and careful in their consideration of evidence, and scientists can occasionally rely on subjective, social, and other criteria to decide between theories. There is, however, an important feature of scientific work that is seldom talked about in philosophical circles, which is the *attitude* that guides scientific inquiry. Even if scientists cannot always rely on a set of rules to follow, it is clear from the history of science that they must rely on *something*. An ethos. A spirit of inquiry. A belief system that tells them that the answer to empirical questions will be found not in deference to authority or ideological commitment—or sometimes even in reason—but in the evidence they gather about the subject matter under investigation. Such a creed, I maintain, is the best way of understanding what is special about science. I will call this the *scientific attitude*.

The scientific attitude can be summed up in a commitment to two principles:

(1) We care about empirical evidence.

(2) We are willing to change our theories in light of new evidence.

This, of course, does not rule out the idea that other factors may sometimes weigh in. As Thomas Kuhn's work demonstrates, even when we care about evidence it can underdetermine theory choice, which opens the door to extra-empirical considerations. What must be ruled out, however, is wishful thinking and dishonesty. In the pithiest version of his ongoing attempt to capture what is distinctive about science, Richard Feynman tells us that "science is what we do to keep from lying to ourselves."[2] Perhaps there is no better rendering of the proper mindset behind the scientific attitude than this.

Such talk about the attitudes and values of science may be dismissed by some as too vague and unrigorous to be helpful, so let me now be more specific about what this approach might entail. What does it mean to care about evidence? Maybe the best way to think about this is to examine what it means *not* to care about evidence. If one does not care about evidence, one is resistant to new ideas. One is dogmatic. Such a person might hold onto their beliefs no matter what the evidence shows. When the scientific attitude says that we must "care about evidence" the idea is that we must be *earnestly willing to seek out and consider evidence that may have a bearing on the grounds for our beliefs*. In some cases this will improve our justification, but in others it may undermine it. Scientists must be open to either.

To care about evidence is to be willing to test our theory against a reality that might refute it. It is to commit to hold a belief not because it makes us feel good, seems right, or even coheres with other things that we believe, but because it fits with the data of our experience. While there is a vast literature in the philosophy of science that shows just how difficult it is sometimes to decide between theories on this basis—necessitating all sorts of other considerations such as simplicity, fruitfulness, and coherence—this does not change the underlying creed of science: where it is available, evidence should drive scientific theory choice.[3]

Of course, with some topics, we might not care about evidence because it is irrelevant. If the subject is math or logic, then evidence will not make a difference because anything at issue can be resolved through reason. But when an empirical subject is under investigation, this kind of rejection of evidence is anathema to rigorous inquiry. In science, we seek knowledge from experience in order to see what the world is like. Caring about evidence is fundamental because it is the only way to shape our knowledge closer and closer to the reality that we seek to know.

One might imagine here a list of traits. The person with a good scientific attitude is humble, earnest, open-minded, intellectually honest, curious, and self-critical. The danger here, however, is that we cannot simply make the scientific attitude a matter of individual psychology, nor can we leave it to the judgment of the individual whether he or she possesses these traits. For one thing, what to do about the denialist or pseudoscientist, who might *say* that they care about evidence—or even actually believe it—when it is obvious to the rest of us that they do not? Such a person may simply be lying to us, but they may also be lying to themselves. If the scientific attitude were just a matter of how one *feels* about whether one cares about evidence, it would not be possible to differentiate between the genuinely earnest person who is searching for a way to test their beliefs against experience versus the ideologues who are deluded into thinking that they care about evidence merely because they cherry pick facts that confirm their prior beliefs. Instead, the scientific attitude must be measured by our actions, and these actions are best judged not by the individual, but by the larger community of scientists, who share the scientific attitude as a guiding ethos.[4] *To care about evidence is thus to act in accordance with a well-vetted set of practices that have been sanctioned by the scientific community because they have historically led to well-justified beliefs.*

This is not to say that the process of science is perfect. Even when we fully embrace it, the scientific attitude probably cannot eliminate all of the denialists and pseudoscientists who say that they have it, even when they do not. Whether they are fooling themselves or others, it is sometimes hard to tell.[5] Likewise, there may be scientific researchers who sometimes get too close to their own theories and refuse to believe what the data tell them.[6] Where is the line between these two camps? Even if the scientific attitude cannot draw a firm logical or methodological division between science and its imposters, it can at least expose a basic lacuna in values that are illuminated by how we behave in the face of contravening evidence.

What is evidence? It is probably impossible to define all of the different things that may count as scientific evidence. Statistical, qualitative, or even historical evidence may exist across different empirical endeavors. Evidence is the data we get from experience that affects our rational degree of belief in a theory. Sometimes this data is quantitative and can be measured directly. Other times it is diffuse and must be interpreted. Either way, scientists should agree that evidence is crucial in choosing or modifying a scientific theory.

There are, however, many competing theories of what it means for scientists to use evidence in a rational way. As Peter Achinstein writes:

> Scientists frequently disagree about whether, or to what extent, some set of data or observational results constitute evidence for a scientific hypothesis. Disagreements may be over empirical matters, such as whether the data or observational results are correct, or whether other relevant empirical information is being ignored. But conflicts also arise because scientists are employing incompatible concepts of evidence.[7]

Entire books have been written about these different concepts of evidence, and their various strengths and weaknesses in explaining how the facts of experience either do or do not lend support to a scientific theory.[8] It might shock those outside the philosophy of probability and statistics to learn that there are competing accounts of what it is appropriate to infer from one and the same piece of evidence. There are furious debates, for instance, over the "subjectivist" approach to probability favored by Bayesians versus the "frequentist" approach offered by Deborah Mayo and others.[9] The scientific attitude may be thought of as compatible with many different concepts of evidence. No matter your theory of the proper way to *use* evidence, the scientific attitude *toward* evidence is one where you are committed to the idea that evidence is paramount in making up your mind about whether a theory is worthy of belief.

Of course, the best way to appreciate the importance of the scientific attitude is to see it in action, and I'll soon give a few examples. First, however, I would like to address two possible misconceptions. First, the scientific attitude is *not* meant to be a solution to the problem of demarcation.[10] The goal of the demarcation project is to find a logical criterion by which one can sort all and only science into one camp, and everything else in the other. That is a tall order and, as we have seen, virtually every attempt to do this has failed. This leaves science open to misunderstanding and criticism by those who do not fully appreciate what it is about. The goal of identifying the scientific attitude as an essential feature of science is not to wall it off from other disciplines but to show that unless those who make empirical claims are willing to follow the rigorous standards that define scientific reasoning, they will fall short of the best way the human mind has ever devised of coming to know the empirical world.

A second possible misconception may involve my intentions. I am not here attempting to give a descriptive account of what scientists actually do because, in any given lab on any given day, one's commitment to the scientific attitude may be in flux. Scientists may occasionally violate the norms of science and then later, one hopes, come back into line.[11] Instead, I offer the scientific attitude as a normative ideal, by which we may judge whether some individual scientist or entire field of inquiry is living up to the values of science. As we've seen, science is not based on some formula, as the scientific method promised. Neither is it strictly a matter of some all-or-nothing judgment about logic or methodology. Science is defined by a set of practices that are embedded in the values upheld by the people who perform it.

This is not to say that science is solely justified (or not) by what it does. Practice is important but it is not the only thing that matters in judging science. I say this because one strain of argument against the methodological approach to the philosophy of science over the last few decades has been that, because science does not always follow the precepts that the logic of science would dictate, science must be no better or worse than any other form of inquiry. I believe this conclusion to be misguided. Although I myself eschew a methodological defense of science, my approach too is rooted in the idea of rational justification. I may not agree with the traditional idea that in order to defend science one must draw a rigid distinction between facts and values, but I do not believe that science is hopelessly subjective either. Although objectivity is important, values play a role by guiding our practice and keeping us on track. This is to say that even though the practice of science may sometimes fall short, it is still possible to justify science as a whole based on the goals of its aspiration.

To recognize the role of practice in understanding science does not diminish the importance of its ideals. Though some may cheat or do sloppy work, this does not mean that science is unjustifiable. Just as it does not undermine the logic of science to say that individual scientists are sometimes irrational, it does not undermine the values of science to point out that some practitioners have occasionally betrayed the scientific attitude. Indeed, this is why it is important to champion the role of group scrutiny in judging scientific work. The standards of science are upheld not just by the individual, but by the community of scientists, who have developed a set of tools to keep it honest. This is why the scientific attitude is a normative rather than a descriptive thesis. Humans sometimes cheat on an ideal even when they believe in it. In such a case, it is up to others to offer correction. And this is exactly what the scientific attitude allows one to do. What makes science distinctive is not merely what it *does*, but what it *aims* to do. Despite any mistakes made by the individual, it is the *ethos* of science that affords it such great epistemic authority.[12]

Two Examples of the Scientific Attitude

I promised at the outset of this book that one could learn the most about science by looking not just at its successes but also at its failures. I also promised not to use examples exclusively from the history of physics and astronomy. Given that, I will now pursue one example that illustrates the virtues of the scientific attitude drawn from medicine, followed by a "failed" example from chemistry (cold fusion), which demonstrates what can happen when the scientific attitude is compromised.[13]

For this strategy to be credible, it must square with the belief that it would have been easy to find other examples from the history of physics and astronomy that would also demonstrate the merits of the scientific attitude. I think this is not an unreasonable assumption. One could, for instance, turn to Newton's theory of gravity or, better yet, to Einstein's general theory of relativity. Given the head start we have from Popper's reliance on this example, this might serve well to illustrate the power of having the proper mindset when one is testing a theory. But I will leave it to the reader to imagine what Popper might have said about the scientific attitude of Einstein. Instead I will here share one of my own favorite examples from the history of science: Semmelweis's theory of childbed fever. This example was made famous within the philosophy of science by Carl Hempel, who used it in his 1966 book *Philosophy of Natural Science* to illustrate the virtues of scientific explanation.[14] For my own purposes, I will take pains to highlight the way that Semmelweis's theory demonstrates the scientific attitude, rather than the logical empiricist account of science within which Hempel frames it. This will also tie in quite well with what I have to say about how the scientific attitude transformed modern medicine in chapter 6.

Given the unquestionably scientific status of modern medicine, it is hard to believe that more than two hundred years after the start of the scientific revolution in the seventeenth century, medical care was still in the dark ages. As late as 1840, medical care still had not enjoyed the discovery of anesthesia (1846), the germ theory of disease (1850s), or antiseptic surgery (1867). One problem is that even when discoveries were made, there were few agreed-upon routes for disseminating the information or overcoming the objections of skeptics.[15] Experimental methods took a back seat to intuition and tradition. This makes it all the more remarkable that in 1846, at the Vienna General Hospital, we see one of the greatest examples of the scientific attitude in full flower.

Ignaz Semmelweis was a lowly assistant physician in the world's largest maternity clinic, which was divided into two wards. In Ward 1, childbed fever (also known as puerperal fever) was rampant and the mortality rate was as high as 29 percent; in the adjacent Ward 2, the rate was only 3 percent.[16] Another piece of relevant information was that women who delivered their babies at home or even on the way to the hospital in a "street birth" had a much lower incidence of childbed fever. What was so different about Ward 1? Various hypotheses were offered. One was that Ward 1 was overcrowded. When Semmelweis counted up the patients, however, he noted that the overcrowding was in fact much worse in Ward 2 (perhaps because of all those women avoiding the notorious Ward 1). It was then noted that, because of the physical layout in Ward 1, the priest who was summoned to give last rites to women who were dying of childbed fever was required to pass by many other beds—all while ringing a bell—which might put great fear into the other women and perhaps increase their chances of contracting childbed fever. In Ward 2, the priest had direct access to the sickroom. Semmelweis decided to try an experiment in which he asked the priest to take a different, silent, route to the sickroom in Ward 1, but the mortality rate from childbed fever stayed the same.

Other tests involving whether the women lay on their sides or their backs while giving birth were similarly fruitless. Finally it was noted that one of the main differences was that in Ward 1 the deliveries were handled by medical students, whereas in Ward 2 they were performed by midwives. Were the medical students giving rougher examinations? After the medical students and the midwives changed places, the mortality rates followed the medical students, but still no one knew why. After instructing the medical students to use gentler techniques, the mortality rate still did not improve.

Eventually, enlightenment came in 1847 when one of Semmelweis's colleagues received a puncture wound during an autopsy on a woman with childbed fever, and died of an illness that appeared to have the same symptoms.[17] Could childbed fever be contracted by someone other than pregnant women? Semmelweis realized that there was a difference in where the medical students were *before* they came to the maternity ward; they came directly from performing autopsies, with unwashed hands and instruments (remember that this was before antisepsis and the germ theory of disease), straight to the maternity ward, leading to the hypothesis that childbed fever may have to do with the transfer of "cadaveric matter" to the pregnant women. As a test, Semmelweis ordered the medical students to wash their hands in chlorinated water before performing their deliveries. The mortality rate plummeted. He now had an explanation not only for why the incidence of childbed fever was so much greater in Ward 1, but also for why "street births" saw such a low incidence of childbed fever. Eventually, Semmelweis was forced to broaden his hypothesis to include the idea that childbed fever could also be transferred from putrified *living* tissue, after he and his colleagues examined a woman with cervical cancer, and then a dozen other women in succession, eleven of whom died of childbed fever.[18]

Use of the scientific attitude in this example is obvious. Semmelweis did not assume that he already knew the answer to the question of what caused childbed fever; he examined the similarities and differences between the two wards, then learned what he could through observation and controlled experiment. He came up with various hypotheses, then began to test them one by one. When a hypothesis flamed out, he moved on to the next one, leaving himself open to learning new information along the way. Finally, when he found the answer—and later broadened it—he changed his ideas based on the new data.

Did he "care about empirical evidence"? Clearly he did. By controlling the circumstances and testing his ideas against actual experience, Semmelweis was respecting the idea that the cause of childbed fever could not be discerned merely by reason. Was he "willing to change his theory based on new evidence"? Again the answer is yes. Not only did Semmelweis change his hypothesis each time one was refuted, he enlarged it when new information came to light that it was not only cadaveric matter—but also putrid living tissue—that could transfer disease from one body to another. He still did not know the exact mechanism of transfer for the disease (much as Darwin did not know about genetics when he proposed his theory of evolution by natural selection), but the correlation was undeniable. Semmelweis had shown that "lack of cleanliness" was responsible for childbed fever.

Incredibly, this idea was resisted and ignored for decades. Despite Semmelweis's incontrovertible empirical demonstration that chlorinated hand washing could radically diminish the incidence of childbed fever, his hypothesis was challenged by the majority of medical practitioners. Countless women unnecessarily lost their lives as the stubborn medical establishment resented the implication that they were the ones who were giving their patients childbed fever. They were insulted by the idea that gentlemen could be seen as somehow unclean. With no explanation for *how* cadaveric matter could be transferring illness, they were reluctant to give up the hypothesis that it was probably the result of "bad air." Semmelweis was fired from his job and, after further demonstrations of the efficacy of his ideas at other hospitals throughout Europe (still with no acknowledgment from the medical community), he became bitter. Eventually he was committed to an asylum, where he was beaten by guards and died two weeks later of sepsis, a blood infection similar to childbed fever.

In the resistance to Semmelweis's hypothesis, we can see the flip side of the scientific attitude as well. It is true not only that the presence of the scientific attitude facilitates progress in scientific discovery and explanation, but that its absence can impede it. During the 1840s, the medieval concept that disease resulted from an imbalance in the body's "four humors" was still widespread. Custom and tradition dictated answers to medical questions more than empirical discovery. It was not until Pasteur's and Koch's work on the germ theory of disease in the 1850s, and Lister's later introduction of antiseptic surgery in 1867, that medicine began to find its scientific footing. Years after Semmelweis's death, his ideas were vindicated.[19]

One could be forgiven for thinking that this is all so simple and obvious that the people who ignored Semmelweis must have been fools. How could they have been so stubborn and ignorant to miss what was right in front of them? The answer is that up until the middle of the nineteenth century, medicine did not embrace the scientific attitude. The idea that we could learn about an empirical subject through careful experimentation and observational evidence had already taken hold in the physical sciences. Galileo's revolution was more than two hundred years old in astronomy. But old ideas held a firm grip on medicine until much later. Indeed, perhaps the most stunning part of the story about childbed fever is not why so many medical practitioners rejected controlled experimentation and learning from empirical evidence, but that Semmelweis ran so far ahead of the pack and embraced it.[20]

But what excuse do we have for some of today's scientists who occasionally pursue research that does not live up to this standard? It is perhaps ironic that one of the most compelling demonstrations of scientists' reliance on the power of evidence can be shown through what some have called the worst example of scientific bungling in the twentieth century. In the spring of 1989, two chemists from the University of Utah—B. Stanley Pons and Martin Fleischmann—held a press conference to announce that they had achieved a sustained nuclear fusion reaction at room temperature. If true, the implications would be enormous, for it would mean that the dream of a clean, cheap, and abundant source of worldwide energy might soon be fulfilled. As expected, scientists met this announcement with enormous skepticism—not least because it was made via press conference rather than the more customary route of publication after rigorous peer review—and set about immediately to try to reproduce Pons and Fleischmann's results.

And they could not. After a two-month honeymoon in the media, during which time other scientists were handicapped by Pons and Fleischmann's refusal to share the details of their experiment, it was shown that their work was hopelessly flawed. Accusations of extrascientific meddling ran wide, but in the end all that mattered was appeal to the evidence. Many scientists were extremely embarrassed by this whole episode, especially when books began to appear with titles such as *Bad Science*, *Too Hot to Handle*, and *The Scientific Fiasco of the Century*. Rather than being ashamed of this, scientists might instead have celebrated this occasion to demonstrate the power of scientific skepticism. Despite all of the money, prestige, and media attention, the case was decided by empirical evidence. Though one particular theory (and a couple of reputations) had been spectacularly shot down, this was a victory for the scientific attitude.[21]

Here one sees a situation that is almost the opposite of what faced Semmelweis. In the case of childbed fever, it was the lone practitioner who insisted that his results were correct, if anyone would bother to look at the evidence. With cold fusion, the original experimenters were perhaps too blinded by the hype surrounding their theory to be more deliberate with their inquiry and release their results only after they had survived a little more methodological self-scrutiny, an attempt at replication, and peer review. Fortunately, with cold fusion, the scientific attitude *was* embraced by the larger scientific community, who acted as a check against the haste and preference for one's own theory that can sometimes derail scientific research. For the wider scientific community—some of whom surely had their own interests at stake—the proper way to decide it was to see what evidence could be brought to bear on the problem.

It is not that mistakes are never made in science. Scientists are human and thus subject to all of the traits of ambition, ego, greed, and stubbornness that motivate the rest of the human population. What is remarkable is that in science we have agreed-upon, transparent standards that can be used to adjudicate empirical disputes and try to correct any errors. For Semmelweis, the medical field waited two decades for the right theory to become firmly entrenched. With cold fusion, it took only two months. The difference was the presence of the scientific attitude.

Roots of the Scientific Attitude

The idea that scientists' attitude is an important feature of science is not new. It has been anticipated by many others, including Popper and Kuhn.[22] Popper, in his account of falsification, emphasized the idea that there is a "critical attitude" behind science. Indeed, in some sense Popper seems to feel that a critical attitude is *prior* to falsifiability.[23]

> What characterizes the scientific approach is a highly critical attitude towards our theories rather than a formal criterion of refutability: only in the light of such a critical attitude and the corresponding critical methodological approach do "refutable" theories retain their refutability.[24]

In his intellectual autobiography, Popper reflects on how he first came to the idea of falsification and draws a connection between the critical attitude and the scientific attitude:

> What impressed me most was Einstein's own clear statement that he would regard his theory as untenable if it should fail in certain tests. Thus he wrote, for example: "If the redshift of spectral lines due to the gravitational potential should not exist, then the general theory of relativity will be untenable." Here was an attitude utterly different from the dogmatic attitude of Marx, Freud, Adler, and even more so that of their followers. Einstein was looking for crucial experiments whose agreement with his predictions would by no means establish his theory; while a disagreement, as he was the first to stress, would show his theory to be untenable. This, I felt, was the true scientific attitude. It was utterly different from the dogmatic attitude which constantly claimed to find "verifications" for its favourite theories. Thus I arrived, by the end of 1919, at the conclusion that the scientific attitude was the critical attitude, which did not look for verifications

> but for crucial tests; tests which could refute the theory tested, although they could never establish it.[25]

I applaud this account. There is an important aspect of the scientific attitude that is captured in Popper's insight about falsificationism. I disagree with Popper, however, that the best way to capture this critical attitude is to reduce it to a methodological principle that serves as a criterion of demarcation. As Popper recognizes, there is something special about the attitude that scientists have toward the power of empirical evidence. But need this be a matter of logic?

Kuhn also recognized the importance of the scientific attitude. This fact is often overlooked, owing to the enamored response that Kuhn's account of science received at the hands of the "Strong Programme" of sociology of science, which argued that all scientific theories—both true and false—could be explained by sociological rather than evidential factors, and were thus in some sense relative to human interests. Kuhn, however, was dismayed by this interpretation of his work and resisted the idea that nature didn't matter to scientists. As one commentator writes:

> Kuhn ... was deeply troubled by the developments in the sociology of science initiated by the Strong Programme. ... Kuhn was concerned that the proponents of the Strong Programme misunderstood the role that values play in science. ... He complained that the Strong Programme's studies of science "leave out the role of [nature]." ... Kuhn, though, insists that nature plays a significant role in shaping scientists' beliefs.[26]

While Kuhn took seriously the idea that theories must be compared to one another, they must also be tested against empirical evidence. Kuhn writes:

> [The world] is not in the least respectful of an observer's wishes and desires; quite capable of providing decisive evidence against invented hypotheses which fail to match its behavior.[27]

Unlike Popper, Kuhn may not have framed this as an "attitude" that stood behind the methodology of science, but Kuhn nonetheless recognized the important role that empirical evidence could play in helping scientists to decide between theories and saw that commitment to the value of empirical evidence was necessary for science to go forward.

One is thus left with the question of why neither Popper nor Kuhn went so far as to make the values of science—either the critical attitude or respect for the idea that nature could overrule our wishes and desires—the basis for distinguishing between science and nonscience.[28] For Kuhn, the answer is perhaps easier: although he felt that science was special, and took great pains to come to terms with how science actually worked, he did not wish to tie himself to any formal criterion of demarcation.[29] Popper, on the other hand, overtly *did* wish to do this, so it is perhaps more of a live question why he did not make more of an attempt to find within the critical attitude of science an explanation for what is distinctive about it. One could argue, I suppose, that he did precisely this through his account of falsification. Yet at some level this fails to come to grips with Popper's strategy of drawing a distinction between the way that science operates versus how philosophers try to justify it, juxtaposed against his deep ambivalence over how to deal with the fact that practical considerations could sometimes threaten the beauty of his logical account of demarcation. Even within one of his clearest statements that falsification is a logical solution to the problem of demarcation, Popper writes: "what is to be called a 'science' and who is to be called a 'scientist' must always remain a matter of convention or decision."[30] Clearly Popper understood the importance of flexibility, having a critical attitude, and occasional deference to practical matters. Still, he yearned to have an absolutely *logical* basis for drawing a distinction between what was science and what was not. I think that this distracted him from recognizing the full power of something like the scientific attitude, which may not have seemed "hard" enough to satisfy his mandate for logical demarcation.

An even deeper root of the scientific attitude may be found at the very beginning of when philosophers started to think about the methodology of science. Although he is primarily remembered today for his work on scientific method, the idea that there are special "virtues" that attach to scientific inquiry can be found in Francis Bacon's 1620 masterpiece *The New Organon*.[31] Here he offers virtues like honesty and openness as inextricably bound up with the good practice of science. Bacon asserts that methodology is important, but it must be embedded within the appropriate values that support it. Indeed, Rose-Mary Sargent has maintained that the modern quest for "objectivity" in the defense of science—where one attempts to bifurcate facts from values—represents a perversion of Bacon's ideas.[32] It may seem ironic that the person most often associated with the idea of scientific method would also uphold the idea that scientific practice must be pursued with the appropriate attitude, but one has only to read the preface and first fifty aphorisms of *The New Organon* to confirm Bacon's intention.

In his subsequent work *The New Atlantis* (1627), Bacon also pushed for the idea that these scientific virtues must be expressed not only by individual practitioners, but also by the community of scientists who would judge and uphold them. In her paper "A Bouquet of Scientific Values," Noretta Koertge recounts the communal nature of Bacon's vision for science. She writes, "Bacon's dream of a new science comprised not only a new methodology but also a community dedicated to the task."[33] Thus we see that a fairly robust account of the scientific attitude has been there all along, practically since the birth of talk about "scientific method."

Finally, the roots of the scientific attitude may perhaps be appreciated by analogy with another philosophical field that can trace its origins all the way back to Aristotle. In Alasdair MacIntyre's classic *After Virtue*, he asks us to consider the merits of his "community practice" approach to normative ethics by way of a chilling thought experiment about science:

> Imagine that the natural sciences were to suffer the effects of a catastrophe. A series of environmental disasters are blamed by the general public on the scientists. Widespread riots occur, laboratories are burnt down, physicists are lynched, books and instruments are destroyed. Finally a Know-Nothing political movement takes power and successfully abolishes science teaching in schools and universities, imprisoning and executing the remaining scientists. Later still there is a reaction against this destructive movement and enlightened people seek to revive science, although they have largely forgotten what it was. But all that they possess are fragments: a knowledge of experiments detached from any knowledge of the theoretical context which gave them significance; parts of theories unrelated either to the other bits and pieces of theory which they possess or to experiment; instruments whose use has been forgotten; half-chapters from books, single pages from articles, not always fully legible because torn and charred. Nonetheless all these fragments are reembodied in a set of practices which go under the revived names of physics, chemistry and biology. Adults argue with each other about the respective merits of relativity theory, evolutionary theory and phlogiston theory, although they possess only a very partial knowledge of each. Children learn by heart the surviving portions of the periodic table and recite as incantations some of the theorems of Euclid. Nobody, or almost nobody, realizes that what they are doing is not natural science in any proper sense at all.[34]

In such a world, what would be missing? Precisely the thing that makes science so special. Even if we had all of the content, knowledge, theories—and even the methods—of science, none of this would make sense without the values, attitudes, and virtues of scientific practice that enabled these discoveries to occur in the first place.

Here the analogy between virtue ethics and science is made explicit.[35] In the great ethical debate that has come down to us since Aristotle, some have argued that what makes right acts right is not their adherence to some normative moral theory that purports to delineate our duties based on how well they conform to an ideal standard about consequences (such as utilitarianism) or adherence to some rational principle (such as deontology); rather what makes moral behavior moral is the virtue of the people who perform it. People with good moral character behave morally; morality is what moral people do.

Can a similar move now be made in the debate over science? I do not believe it is quite so simple to say that science is simply what scientists do; as in the debate over ethics, we have also to consider the nature and origin of our values, and how they are implemented and judged by the wider community.[36] Nonetheless, the analogy is intriguing: maybe we need to focus less on demarcating scientific from nonscientific *theories* and more on the virtuous epistemic *attitudes* that are behind the practices of science.[37] Some of this work is just beginning in the field of virtue epistemology, which proceeds by analogy with virtue ethics: if we want to know whether a belief is justified, perhaps we would do well to focus at least some of our attention on the character, norms, and values of the people who hold it. The application of this to problems in the philosophy of science is still quite new, but there has already been some excellent work in applying the insights of virtue epistemology to such thorny problems as underdetermination and theory choice in the philosophy of science.[38]

I hope it is clear by now that I do not seek to make a priority claim for the scientific attitude. This idea has a long history that goes back through Popper and Kuhn, at least to Francis Bacon, and arguably to Aristotle. What I hope to emphasize is that the scientific attitude has been woefully neglected in the philosophy of science. It can nonetheless play a crucial role in understanding and defending science by illuminating an essential feature that has been missing from many contemporary accounts. If we focus exclusively on method, we may miss what science is most essentially about.

Conclusion

In previous accounts, some philosophers of science have felt that the best way to defend science is to come up with a logical justification of its method, rather than look at how science is actually done. When this resulted in a criterion of demarcation that purported to sort all and only science to one side of the ledger—and all and only that which was not science to the other—trouble ensued. For this reason, it seems a virtue of the scientific attitude that it is flexible enough to capture why scientific explanation is distinctive, yet robust enough to ensure that even without a decision procedure, we can still tell whether inquiry is scientific. Having a scientific "attitude" toward evidence may seem soft, yet it captures the essence of what it means to be scientific. Whether this is an observation about the context of discovery (how science actually works) or the context of justification (a rational reconstruction after the fact) matters little. For I think that the greatest threat to science's credibility comes not from some philosophical distinction between the way that scientists do their work and the method we use to justify it, but instead from the improper introduction of ideological commitments into the scientific process. And the scientific attitude is a bulwark against precisely this sort of ideological infection.[39]

The idea behind the scientific attitude is simple to formulate but difficult to measure. It nonetheless plays a crucial role both in explaining how science operates and in justifying the uniqueness of science as a way of knowing. Science is successful precisely because it embraces an honest and critical attitude toward evidence (and has created a set of practices like peer review, publication, and reproducibility to institutionalize this attitude).[40] Of course, science is not always successful. One can have the scientific attitude and still offer a flawed theory. But the power of caring about empirical evidence is that we (and others) may critique our theory and offer a better one. When we are trying to learn about the empirical world, evidence must overrule other considerations.[41] The evidence may not always be definitive, but it cannot be ignored, for the check that it gives us against reality is the best means of discovering (or at least working toward) the truth about the world.

This highlights again the tentative nature of any scientific theory. The scientific attitude is fully consonant with the idea that we can never be sure that we have the truth: all theories are provisional. But this is as it should be, for what is distinctive about science isn't the truth of Newtonian or

Einsteinian theory, but the process by which those theories came to be warranted. Scientific theories are believable not merely because they fit with the data of our experience, but because they are built through a process that respects sensory evidence and the idea that our untutored hypotheses can be improved by a clash with the searching criticism of other scientists who have seen the same data.

We are reminded again of the fragility of science, for it depends on the willingness of its practitioners to embrace the scientific attitude. No matter how reliable our method, science could not work without the forthright and cooperative spirit of scientists. If we want to be scientists, our commitment cannot be to any given theory (even our own) or any ideology, but must be to the scientific attitude itself. Other factors may count, but in the end they are and always should be trumped by evidence. While this may sound meager, it is the heart of what is distinctive about science.

4 The Scientific Attitude Need Not Solve the Problem of Demarcation

If the scientific attitude is our best means of articulating what is special about science, the question inevitably arises: could it be the long-awaited solution to the problem of demarcation? If so, would it need to provide a set of necessary and sufficient conditions for telling the difference between science and nonscience? It is my contention that the philosophy of science need not get sidetracked into this issue, when all we really need to do is identify one simple necessary condition for good science: caring about evidence and being willing to use that evidence to revise one's theories. If some potential area of inquiry fails to do this, then it is not science. Thus, in keeping with the theme of this book, one sees that we can learn quite a bit about science by looking at what it is *not*. One need not prove that anything *with* the scientific attitude *is* science; one need only show that anything *without* the scientific attitude is *not*. Whether this constitutes a full-blooded criterion of demarcation in the most robust sense of the word seems less important than that it gives us what we need to understand what is most distinctive about science.

A Short Detour through the Vexed Question of Necessary and Sufficient Conditions

Recall here Laudan's admonition: without specifying a set of necessary and sufficient conditions for science, one cannot hope to solve the problem of demarcation. Yet given the repeated failure of philosophers of science to do this, he argues, the demarcation problem is dead. Of course, even if Laudan is right, this does not mean that one could have no fruitful discussion over whether there is something special about science. But it is also possible that Laudan is incorrect in saying that providing a set of necessary and sufficient conditions is itself necessary (or sufficient) for solving the problem of demarcation.[1]

The key insight here is to realize just how high one is putting the bar to say that we have to provide a set of necessary and sufficient conditions. In saying that A is *necessary* for B we are saying "if B, then A." Likewise, in saying that A is *sufficient* for B we are saying "if A, then B." Any student of logic will recognize the equivalences at work here and understand that saying "A is sufficient for B" is equivalent to saying "B is necessary for A," and likewise that "B is sufficient for A" is equivalent to saying "A is necessary for B." But this is where the magic happens, for when you combine these two statements into one that says "A is both necessary and sufficient for B"—which of course means the same thing as "B is both necessary and sufficient for A"—you get the logical equivalence of A and B. And this is the strongest relationship one can get in logic.[2]

This sets a formidable task for demarcation and the implications are staggering, for it means that in searching for the necessary and sufficient conditions for science we are searching for some criterion that is *logically equivalent* to science. The best illustration of the cost of this comes from Popper. Recall that Popper equivocates over whether his notion of falsifiability—as a solution to the problem of demarcation—is intended to provide a necessary condition for science, a sufficient one, or one that is both necessary and sufficient.[3] Laudan takes Popper to task for his sufficiency condition, arguing that it does not do enough to maintain the high standards of science: What should we do about fields like astrology that make demonstrably false claims (which are surely therefore also falsifiable)? Must we admit them as science? Likewise, others have questioned the necessity interpretation of Popper's criterion, arguing that it is too strict: What should we do about scientific fields that seem not to make any falsifiable claims?[4] Are they not sciences? But I would now like to point out that even if Popper *did* offer falsification as both a necessary and sufficient criterion, he is still in hot water. For later in life, when Popper wrote "a sentence (or theory) is empirical-scientific if and only if it is falsifiable,"[5] he is not solving the problems above, but merely combining them. By committing himself to the logical equivalence of science and falsifiability, he suffers criticism both from those who think his account is too strict and from those who think it is too liberal.[6]

To explain this problem more precisely, one must understand the power of the biconditional relationship in logic. To say "A if and only if B" is

Table 4.1

Criterion	Type	Problem
If science, then falsifiable (If not falsifiable, then not science)	necessity	Is evolutionary biology not science?
If falsifiable, then science (If not science, then not falsifiable)	sufficiency	Is astrology science?
Science iff falsifiable (Falsifiable if and only if science)	equivalence	Is astrology science, but evolutionary biology is not?

equivalent to saying "B if and only if A." So to say that "a theory is scientific if and only if it is falsifiable" is equivalent to saying "a theory is falsifiable if and only if it is scientific." But this means that "being falsifiable" and "being scientific" are logically equivalent. Once you put "if and only if" into the sentence, "science" and "falsifiability" have identical truth conditions.[7] (See table 4.1.) A necessity standard is too strong and a sufficiency standard is too weak. But by adding them together, instead of creating a criterion that is "just right," the problems seem to multiply. Now the standard cannot satisfy our intuition that evolutionary biology is scientific but astrology is not.[8]

This is why I renounce the "necessary and sufficient conditions" approach when discussing the scientific attitude as a means of understanding what is special about science. For if one were to try to put the scientific attitude forward as a set of necessary and sufficient conditions, I think that the same type of problems would arise as we just saw with Popper. In particular, defending the distinctiveness of science does *not* seem to require a sufficiency condition. It is enough to embrace the necessity condition and say "For an area of inquiry to be a science, it must have the scientific attitude," which is logically equivalent to saying "If a theory does *not* have the scientific attitude, then it is *not* science." (See table 4.2.)

Of course, I realize that by choosing not to specify a sufficiency condition we will have only identified what is *not* science, not what is. But perhaps that is all one needs. We won't be able to say definitively whether something like string theory is scientific. Or even the rest of physics. But is that really so problematic? Although we may have started out wondering whether there was some way to show why *science* was scientific, perhaps that is the wrong question. Maybe the goal of saying what is distinctive about science is instead to show why some areas of inquiry are *not* scientific. We may define science by its absence. This allows us to protect

Table 4.2

Criterion	Type
If science, then scientific attitude (If not scientific attitude, then not science)	necessity
If scientific attitude, then science (If not science, then not scientific attitude)	sufficiency
Science iff scientific attitude (Scientific attitude iff science)	equivalence

it from imposters. When we discover the essential properties of science, we can use them to say that if something does not have them then it cannot be science. But this does not mean that everything that has the scientific attitude is a science. It is a necessity standard, not a sufficiency one.

Ever since Laudan, those who have tried to provide a criterion of demarcation have run aground on the issue of trying to provide both necessary and sufficient conditions. Not only can it probably not be done (witness the history of the problem both before and since Laudan's "obituary"), but there are serious costs in trying to follow Laudan's mandate. Instead, I propose that we can get the job done by looking only for the necessary conditions for science, one of which is the scientific attitude. Once we have said that if we have science, then we must have the scientific attitude—and have thus found a necessary condition for science—we have also said that if we do *not* have the scientific attitude, then we do *not* have science—which turns out to be the sufficiency condition for *non*science. When we say "If an area of inquiry does not have the scientific attitude, then it is not science," the logical form is "If not SA, then not science," which by contraposition is equivalent to "If science, then SA." (See table 4.3.)

One must be careful here. It is tempting at this point to want to complete the second half of the job and try to mold the scientific attitude into a sufficiency standard for science, so that we can solve the traditional problem of demarcation. But we must avoid this. While the equivalences outlined in table 4.3 are correct, the tasks in case 1 and case 2 are entirely separate. I contend that case 1 is all we need to identify what is special about science. (And of course, as logicians know, case 1 does not in any way imply case 2.) There would be a high cost for engaging case 2 and trying to put forth the scientific attitude as a sufficiency standard for science. Not only would this pull us back into Laudan's quagmire of specifying the necessary and

Table 4.3

Case 1	
If science, then SA	Necessity standard for science
is equivalent to	is equivalent to
If not SA, then not science	Sufficiency standard for nonscience
Case 2	
If SA, then science	Sufficiency standard for science
is equivalent to	is equivalent to
If not science, then not SA	Necessity standard for nonscience

sufficient conditions for science (which would mean defending the assertion that the scientific attitude is logically *equivalent* to science), it would also entangle us in the vexed question of *how many different ways something can fail to be scientific.*

Remember that the sufficiency standard for science is the necessity standard for nonscience. But why should we care at all about the necessity standard for nonscience? There are myriad ways for an inquiry to fail to be scientific. Art and literature are not scientific, primarily because they are not trying to be. Astrology and creationism are a different story; here their subjects of inquiry *do* seem to be empirical, so we should care very much about how they fail to be scientific. All of these fields may lack the scientific attitude, but are we really ready to say that this is a *necessary* condition for *non*science? All one has to say is that if these fields do not embrace the scientific attitude—which none of them do, for different reasons—they are not sciences. Our current project requires only that lack of the scientific attitude is *sufficient* for being nonscience, not that there is only *one way* of being nonscientific. This may be more of a live question when we are interested in some of those fields in social science that want to be considered as sciences but are not yet up to the standard. Yet here again our original insight pays off: there may be many ways of being nonscience, but if one wants to become scientific there is only one way: it is necessary to embrace the scientific attitude. Thus I submit that one can say something interesting about what is special about science without solving the traditional problem of demarcation.

It is important here to be sensitive to the issue of condescension in judging what is and is not science. In contrast to some who have wrestled with the problem of what is special about science, I maintain that the scientific attitude need not force us to say that all of nonscience is an inferior human pursuit. (Indeed, it had better not be, because I would count philosophy in that camp!) Literature, art, music—all are nonscience. They do not have the scientific attitude, but that is completely expected. The problem instead comes from some nonscientific disciplines that *are* trying to present themselves as scientific. With astrology, creationism, faith healing, dowsing, and the like, we encounter disciplines that *are* making empirical claims but refuse to abide by the practices of good science, one of which is that they have the scientific attitude toward evidence. They are, in short, only pretending to be scientific. Here it seems more appropriate to scorn those fields that are masquerading as sciences or claim to have some special access to empirical knowledge outside the channels of scientific practice. When one embarks upon the pursuit of empirical knowledge, it is a serious problem to eschew the values of science.

Can One Still Try to Demarcate Science From Pseudoscience?

Some may object that all I have really done thus far is push the problem of demarcation back a step. For some have approached the problem of demarcation not as a dividing line between science and nonscience, but between science and *pseudo*science. The virtue of this approach is that it goes directly to where the problem lies: those fields that are making empirical claims but only pretending to be sciences. So should I try to use the scientific attitude as a criterion for demarcating between science and *pseudoscience* and just cut off the whole question of the other nonsciences like literature and philosophy that are not making empirical claims? Perhaps some might wish to do this, by adding criteria to the scientific attitude such as "is making empirical claims" and "is not pretending to be scientific"[9] in search of an attempt to solve the problem of demarcation between science and pseudoscience. But I think there are deep problems in proceeding this way.

As noted in chapter 1, if we look at the literature on demarcation we find a troubling equivocation on the issue of whether science is supposed to be juxtaposed with nonscience or with pseudoscience. The Logical Positivists seemed to be concerned with distinguishing science from metaphysics (which one might think of as a type of nonscience that the positivists wished to dismiss with some prejudice). Karl Popper, in his *Logic of Scientific Discovery*, seems to be concerned with nonscience, but by the time he writes *Conjectures and Refutations*, he characterizes his enemy as "pseudoscience."

Even Laudan equivocates on whether it is nonscience or pseudoscience that is the nemesis of science.[10]

Yet this makes a difference. Given the logical issues we have just explored surrounding necessary and sufficient conditions, it is imperative to understand science by looking at *all* of the things that are not science, not just those that are pretending to be. Yes, it is tempting to reserve special opprobrium for the frauds and denialists, the poseurs and the charlatans, and in chapters 7 and 8 I will do precisely this. But in considering the question of demarcation vis à vis the provision of necessary and sufficient conditions, the logic is unyielding: every inquiry must either be science or nonscience. The necessity conditions for science are the sufficiency conditions for *non*science, not *pseudo*science. While it is true that a good deal of contemporary work on the demarcation problem since Laudan interprets the demarcation debate as between science and *pseudo*science, this is problematic. As I argued previously (in chapter 1), both Hansson's and Boudry's accounts founder on precisely this failure to distinguish between nonscience and pseudoscience.[11]

I believe the correct distinction is this: the problem of demarcation properly concerns science versus nonscience. The category of nonscience includes—among other things—domains that are *un*scientific (such as math, philosophy, logic, literature, and art) that do not wish to make empirical claims, and those that are *pseudo*scientific (such as astrology, intelligent design, faith healing, and ESP) that *do* wish to play in the empirical sandbox, even while they flout the standards of good evidence.[12] Other fields, such as ethics and religion, are harder to categorize. Some scholars have held, for instance, that ethics is completely normative and thus should join the rest of philosophy in the "unscientific" category. Others contend that parts of ethics make pretensions to science—or even fulfill this desideratum—so its status as either science or pseudoscience depends on your point of view. Religion, insofar as it makes empirical claims on spiritual grounds, is easier to classify as pseudoscience, yet some contend—as Galileo did—that religion has no business in the empirical realm and so should be considered unscientific.[13] Much as I identify with the desire to heap scorn on the deficiencies and dishonesty of pseudoscience, a more straightforward approach is to recognize that the scientific attitude does its job *both* for identifying why literature and art are not science (because they care about empirical evidence only to the extent that it underlies their basis for creative expression and not to prove or disprove scientific theories) and why astrology and creationism are not science (because they are only *pretending* to care about empirical evidence and are unwilling to revise their theories). It is two birds with one stone. We do not need more than this.

Should I have confined myself in this book to discussing only those fields that sought to make empirical claims and thus attempt a demarcation between science and pseudoscience? For the reasons already given, I have chosen the logically cleaner route. Why search for additional necessary criteria for science (at best) or sufficient criteria (at worst), when the scientific attitude already does the job of specifying what is special about science? Thus I do not think that I have merely pushed the problem of demarcation back a step. I never promised to demarcate between unscientific fields and pseudoscientific ones—between literature and creationism, say.[14] It also seems a virtue of my approach that it allows a path for those nonscientific fields that are *not* pseudoscientific (yet perhaps even for these) to emulate the high standards of science.

This book not only *defends* science, it *proselytizes* for it. Thus, I seek to make room for those fields that may genuinely care about empirical evidence—like the social sciences—but may not have always been true to the scientific attitude, perhaps because they did not realize that this was the essential ingredient in becoming more rigorous. I also believe that embracing the scientific attitude may help to turn a once pseudoscientific field into a successful science, much as occurred in the history of medicine.[15] To understand what is special about science one need not scold or condemn all that is nonscience. To say that qualitative sociology is nonscience is scorn enough; one does not have to lump it in with witchcraft. In reading Popper, one sees him take delight in dismissing the work of Adler, Freud, and Marx in the same breath as astrology. Yet perhaps Freud, for instance, just did not recognize the empirical standards that his inquiry was required to live up to and could perhaps emulate as he sought to become more scientific. If philosophers of science took more care in identifying the necessary conditions for science, could even Freudian psychology be rehabilitated into embracing the scientific attitude?[16] Insofar as possible, I want to remain nonjudgmental about those fields that are nonscientific. The scientific attitude is a standard on the empirical battlefield that may be picked up by any discipline that wants to become more rigorous.

Of course, many questions remain. In chapter 5, I will deal with the question of who judges whether someone has the scientific attitude and against what standard. In chapter 8, I will discuss how to handle those who fail to achieve the scientific attitude, either deliberately or in the false belief that they already embrace it, even though it is evident to the scientific community that something is amiss. At present, however, it is important to deal with two remaining questions that will be on the mind of the demarcationist. First, would it be so terrible if I tried to turn the scientific attitude into a sufficiency standard in addition to a necessary one? And second, even if I am not willing to do this, couldn't I still make a claim to have solved the problem of demarcation by redefining it?

Should Everyday Inquiry Count as Science?

We have already dealt with the fallout from refusing to use the scientific attitude to demarcate between science and pseudoscience, and the consequence that this lumps some disciplines like literature and philosophy in with astrology and creationism on the nonscientific side of the ledger. Earlier we faced the fact that a necessity standard alone will not allow us to say that any given discipline *is* science, only that some are *not*. When we refuse to specify a sufficiency standard for science, this is the price we pay: does reliance on the scientific attitude solely as a necessary standard force us to rule out too much? Call this the problem of *exclusivity*. But what about the other half of the problem? What if we were suddenly tempted to try to offer the scientific attitude as a sufficiency standard—in addition to a necessary one—and thus take a shot at solving the problem of demarcation? This would create problems of its own. Remember Popper: by offering falsifiability as both a necessary and sufficient criterion for science, he did not dissolve his problems but doubled them. The same thing would happen here. If we said that having the scientific attitude was sufficient for being scientific, we would run the risk of permitting all sorts of random empirical inquiries into the pantheon of science. Would demarcationists be happy to consider plumbing and TV repair as sciences? Call this the problem of *inclusivity*.

One salient example—which illuminates those mundane occasions when we may be thinking scientifically but arguably not "doing science"–is looking for one's keys.[17] Suppose that my brother has lost his keys. He knows that he last had them in the house, so he does not bother searching the yard or going out to his car. He had them when he unlocked the front door, so they must be inside the house somewhere. First, he searches all the most likely places—his pants pockets, in the hall drawer, next to the television set—and then engages in an ever-widening search whereby he retraces his steps since he came in the front door. He starts in the hallway, goes to the bathroom where he washed his hands, and then into the kitchen where he made a sandwich before walking to the sofa to sit down and eat. There he finds his keys between the sofa cushions. It could be argued that in this activity he was employing the scientific attitude. He cared about empirical evidence and was attempting to learn from observation (as he looked in various locations and ruled them out one by one). He was also continually changing his hypothesis ("the key might be in X location") as he went. But is this science?

While some may be tempted to say yes, I suspect that most scientists and philosophers of science (and virtually all demarcationists) would want to say no and begin to look for some additional criterion to exclude it. This is the gravitational pull of the sufficiency standard. For how could it be that looking for one's keys, or doing the plumbing, or fixing the television, could count as science alongside physics?[18] One avenue here might be to introduce the earlier idea of "needing to have a theory" as a criterion for doing science. As we saw in chapter 2, this was the downfall of the famous Bode's law in the history of astronomy. It fit the evidence, but only because of massive data manipulation and a bit of luck. For one to "care about evidence and be willing to change one's *theory* based on new evidence," mustn't there be a *theory* at stake? But this is a fraught question. Arguably there *is* a theory at stake for my brother. It is not on par perhaps with the theory of evolution by natural selection, but it may amount to a theory nonetheless. Do I really want at this point to try to demarcate between a genuine and nongenuine scientific *theory*? (See table 4.4.)

This is precisely why I argued earlier that we need to avoid the temptation to offer the scientific attitude as a sufficiency condition for science. If one balked earlier at the idea of not being able to say that even physics was

Table 4.4

Criterion	Type	Problem
If science, then SA (If not SA, then not science)	necessity	Exclusivity: should literature be lumped with creationism?
If SA, then science (If not science, then not SA)	sufficiency	Inclusivity: is "looking for one's keys" science?

scientific, would one be willing now to admit that looking for one's keys *is* scientific? As we saw earlier, Laudan contended that we must solve both halves of the demarcation problem, otherwise we will be open to either the exclusivity or the inclusivity problem. As a philosopher of science, one's first reflex may be to turn to logic and methodology, and try to do this; to find some longer list of criteria by which we might make physics and looking for one's keys come out on different sides of the equation, just as we might hope to do with literature and creationism. But I am not going to do this, because I think it is a vexed question to try to specify the myriad ways that something can fail to be science. The scientific attitude cannot be offered as a sufficiency condition for science, because it will not work as a necessity condition for nonscience. So what does demarcation come down to, besides saying that we already know what we take to be science and what we do not?[19]

The scientific attitude is a powerful idea for identifying what is essential about science, yet it may not be able to sort all of the disciplines so that all and only physics, chemistry, biology, and their brethren end up on the right side of some criterion of demarcation. Looking for one's keys in a systematic way is a mindset. It is an attitude. Whether one is willing to accept this as scientific or not perhaps matters very little. There is no need to create a full set of necessary and sufficient conditions for science just to keep it out. Indeed, by using the scientific attitude as a sufficiency standard, this may be the very thing that allows it to creep in![20] There is virtue in keeping the definition of the scientific attitude simple. It is a disposition, not a procedure. There is no checklist by which we may come to know it; we feel it most by its absence. I therefore resist the idea that we should now try to pile on more and more individually necessary conditions—so that one may end up with a complete and correct set of criteria or even a dreaded five-step method—and instead recognize that the scientific attitude merely identifies a mindset that all good scientists should have toward data. The point of identifying the scientific attitude is not to come up with a laundry list of conditions for science, but rather to point out that this simple essential property is customarily missing from those ideologies that are merely masquerading as science, and also from those other areas of inquiry—like the social sciences—that might wish to step out from the shadow of ideology toward more scientific rigor.[21] To say that the scientific attitude is *necessary* for science seems enough.

In keeping with Emile Durkheim's famous dictum that "The sociologist [should] put himself in the same state of mind as the physicist … when he probes into a still unexplored region of the scientific domain.... He must be prepared for discoveries which will surprise and disturb him,"[22] good scientists should approach their studies with curiosity and humility. They certainly should not feel that they have all the answers before they have examined the evidence (and perhaps not even then). More flexible than falsifiability, the scientific attitude may help us not only to grow science in other areas where there are empirical facts—such as the study of human behavior—but also to criticize those cases of fraud and deception where natural scientists have committed the sin of massaging their data to try to fit their extra-empirical beliefs. And, as we shall see in later chapters, it will help us to identify what is wrong with scientific denialism and pseudoscience as well.

Perhaps some future philosopher of science may wish to search for other necessary conditions that, when added to the scientific attitude, might make up a full set of individually necessary and jointly sufficient criteria by which to rule *in* medicine, physics, evolutionary biology, and string theory and rule *out* astrology, faith healing, "looking for one's keys," and Bode's law.[23] But that is not my task here. All I am trying to do is identify *one very important condition that is essential for science to go forward,* which I have argued is the scientific attitude. This is what is missing in nonscience. Thus I hope to have found a way to defend science by specifying what is most special about it, without getting dragged into providing a set of necessary and sufficient criteria in order to solve the problem of demarcation.

Couldn't the Scientific Attitude Nonetheless Work as a Modified Criterion of Demarcation?

There remains one important question. Even if I reject the idea of finding a set of necessary and sufficient conditions for science, might I try to reclaim the problem of demarcation from Laudan?[24] Many commentators in this debate have noted that neither scientists nor philosophers seem to have much trouble identifying what is science and what is not, even though there is much disagreement over how to do so.[25] As Laudan notes, virtually every attempt to offer a set of necessary and sufficient conditions has failed. Is that just the price one pays for trying to mark the difference between science and nonscience? Or can one perhaps try to demarcate science from nonscience without buying into Laudan's mandate about providing necessary and sufficient conditions?[26]

The strategy above is pursued by Pigliucci in his essay "The Demarcation Problem: A (Belated) Response to Laudan." Here he agrees with Laudan that no one can solve the problem of providing the necessary and sufficient conditions for science, but goes on to say that one can probably solve the problem of demarcation in another way, by acknowledging that there is a "cluster of concepts" at work in science that mark off some "fuzzy boundaries." Recall here our discussion in chapter 1 of Pigliucci's idea that we are better off pursuing a Wittgensteinian project of identifying the "family resemblances" at work rather than any hard and fast logical criteria.[27] This may open the door to a new path for understanding science. But I worry that it may also lead us perilously close to emptying the demarcation debate of any content all together.

In a later essay, "Scientism and Pseudoscience: In Defense of Demarcation Projects," Pigliucci again offers his Wittgensteinian interpretation of a "family resemblance" solution to the demarcation problem (which he now characterizes as the "consensus" view among philosophers of science).[28] Here, however, he seems to have grown irritated with the idea that the sort of "fuzzy boundaries" one must tolerate in the absence of a hard and fast logical criterion should ever admit the kind of "everyday inquiry" we just considered in the last section (such as looking for one's keys) as science. He writes:

> [It] makes little sense to say that plumbing is science, or even that mathematics is science. Science is what scientists do, and scientists nowadays have fairly clearly defined roles, tools, and *modus operandi*, all of which easily distinguish them from plumbers and mathematicians, not to mention philosophers, historians, literary critics and artists.[29]

We seem to have jumped here directly from Hansson's notion that "we know what science is even if we have trouble saying it" to Pigliucci's "science is what scientists do." Yet this seems no criterion at all.[30] Indeed, if one believes *this*, can't one just say "I know science when I see it" and be done with it? If so, one imagines that the creationists and climate change deniers—who have recently adopted some of the relativist tactics of postmodernism—would be overjoyed.[31]

While I agree with Pigliucci that it is best to disavow Laudan's necessary and sufficient conditions approach, I nonetheless think that it is possible to be specific about the essential properties of science. I recognize that—like me—Pigliucci wants to save science from the charlatans,[32] but perhaps he goes too far with his "fuzzy boundaries." Can't one at least have a necessary criterion?

I do *not* believe that science is merely "what is done by scientists," even if it *is* intimately related to the good practices that grow out of the critical values that are shared both by individual practitioners and the larger scientific community.[33] As we will see in the next chapter, the scientific community plays a key role in creating the practical means for policing the beliefs, norms, values, and behavior that make up the scientific attitude. Good scientific practice probably cannot be specified by a complete set of rules, but it is surely more than just what the community of scientists happens to believe.[34] Pigliucci is insightful when he says:

> Science is an inherently social activity, dynamically circumscribed by its methods, its subject matters, its social customs (including peer review, grants, etc.), and its institutional role (inside and outside of the academy, governmental agencies, the private sector).[35]

But I think he is wrong to draw from this the conclusion that science is what scientists do. Has he forgotten that the real threat here is posed by pseudoscientists? I think it is possible to be much more specific and say that what is distinctive about science is the scientific attitude, and that this can be defended by renouncing Laudan only in part, recognizing the virtue of being able to say what is *not* science, even if one cannot definitively say what *is*.

Whether one chooses to see this as complete abandonment of both a necessary and sufficient conditions approach *and* the problem of demarcation—or instead one where, like Pigliucci, I attempt to rescue the problem of demarcation from Laudan's elevated standard—makes little difference to me. There would surely be virtues to accepting my "necessity standards of science and sufficiency standards of nonscience" approach as a solution to some modified problem of demarcation, but there would probably be costs as well. For one thing, how can one use something as amorphous as the beliefs and values that make up the scientific attitude as a criterion of demarcation? How can this be measured? This is why the methodological approach works so well for a criterion of demarcation, but the attitudinal approach perhaps falls short. Yet the pity here seems less for the scientific attitude than for those who seek to explain science solely by fitting it into the framework of demarcation. Attitudes and values are real, even if they are hard to measure. I am thus satisfied with saying that the scientific attitude approach to understanding science may have to give up on the traditional problem of demarcation, even while it retains the idea that one can specify an essential property of science that makes it epistemically privileged.

What would this mean for sorting out the disciplines? Although not a criterion of demarcation, it is perhaps interesting to note how the universe of inquiry might look according to the necessity standard of the scientific attitude. Note that this is *not* a sorting between science and nonscience. With only a necessity standard, one can say only "If it's science, then it has the scientific attitude." One cannot say "If it has the scientific attitude, then it is science." Thus the scientific attitude allows us to be specific only about what is *not* science, not what *is*.[36] (See table 4.5.)

The scientific attitude may offer something less than a traditional criterion of demarcation, but it is a powerful tool in our quest to understand what is special about science. We may not be able to guarantee that every field that *has* the scientific attitude *is* a science, but by showing that every field that does *not* have it is *not* a science, we may yet come to know science more intimately.

But who should judge the crucial question of whether someone is following good evidential standards and so embracing the scientific attitude? As we will see in the next chapter, this goes beyond the values of any individual scientist and must involve the judgment of the entire scientific community.

Table 4.5

Has the Scientific Attitude
Natural science (beliefs based on evidence)
Areas of legitimate scientific disagreement (evidence not clear, so withhold belief)
Wrong, but scientific (false theory, but warranted given evidence at the time)
Social sciences (that use experimentation and evidence)
Everyday inquiry (that is evidence based)
Does Not Have the Scientific Attitude (Nonscience)
Math/logic (not empirical)
Literature, art, and the like (not empirical; not trying to be science)
Social sciences (that are not evidence based)
Bad science (sloppy, mistakes, cut corners)
Fraud (lies, cheats, misleads)
Pseudoscience (pretend to be science, but refuse to embrace good evidential standards)
Denialism (ideology based; don't care about evidence)

5 Practical Ways in Which Scientists Embrace the Scientific Attitude

For science to work, it has to depend on more than just the honesty of its individual practitioners. Though outright fraud is rare, there are many ways in which scientists can cheat, lie, fudge, make mistakes, or otherwise fall victim to the sorts of cognitive biases we all share that—if left unchallenged—could undermine scientific credibility.

Fortunately, there are protections against this, for science is not just an individual quest for knowledge but a group activity in which widely accepted community standards are used to evaluate scientific claims. Science is conducted in a public forum, and one of its most distinctive features is that there is an ideal set of rules, which is agreed on in advance, to root out error and bias. Thus the scientific attitude is instantiated not just in the hearts and minds of individual scientists, but in the community of scientists as a whole.

It is of course a cliché to say that science is different today than it was two hundred years ago—that scientists these days are much more likely to work in teams, to collaborate, and to seek out the opinion of their peers as they are formulating their theories. This in and of itself is taken by some to mark off a crucial difference with pseudoscience.[1] There is a world of difference between seeking validation from those who already agree with you and "bouncing an idea" off a professional colleague who is expected to critique it.[2] But beyond this, one also expects these days that any theory will receive scrutiny from the scientific community writ large—beyond one's collaborators or colleagues—who will have a hand in evaluating and critiquing that theory before it can be shared more widely.

The practices of science—careful statistical method, peer review before publication, and data sharing and replication—are well known and will be discussed later in this chapter. First, however, it is important to identify and explore the types of errors that the scientific attitude is meant to guard against. At the individual level at least, problems can result from several possible sources: intentional errors, lazy or sloppy procedure, or unintentional errors that may result from unconscious cognitive bias.

Three Sources of Scientific Error

The first and most egregious type of error that can occur in science is intentional. One thinks here of the rare but troubling instances of scientific fraud. In chapter 7, I will examine a few of the most well-known recent cases of scientific fraud—Marc Hauser's work on animal cognition and Andrew Wakefield's work on vaccines—but for now it is important to point out that the term "fraud" is sometimes used to cover a multitude of sins. Presenting results that are unreproducible is not necessarily indicative of fraud, though it may be the first sign that something is amiss. Data fabrication or lying about evidence is more solidly in the wheelhouse of someone who is seeking to mislead. For all the transparency of scientific standards, however, it is sometimes difficult to tell whether any given example should be classified as fraud or merely sloppy procedure. Though there is no bright line between intentional deception and willful ignorance, the good news is that the standards of science are high enough that it is rare for a confabulated theory—whatever the source of error—to make it through to publication.[3] The instantiation of the scientific attitude at the group level is a robust (though not perfect) check against intentional deception.

The second type of error that occurs in science is the result of sloppiness or laziness—though sometimes even this can be motivated by ideological or psychological factors, at either a conscious or an unconscious level. One wants one's own theory to be true. The rewards of publication are great. Sometimes there is career pressure to favor a particular result or just to find something that is worth reporting. Again, this can cover a number of sins:

(1) Cherry picking data (choosing the results that are most likely to make one's case)

(2) Curve fitting (manipulating a set of variables until they fit a desired curve)

(3) Keeping an experiment open until the desired result is found

(4) Excluding data that don't fit (the flip side of cherry picking)

(5) Using a small data set

(6) P-hacking (sifting through a mountain of data until one finds some statistically significant correlation, whether it makes sense or not)[4]

Each of these techniques is a recognizable offense against good statistical method, but it would be rash to claim that all of them constitute fraud. This does not mean that criticism of these techniques should not be severe, yet one must always consider the question of what might have motivated a scientist in any particular instance. Willful ignorance seems worse than carelessness, but the difference between an intentional and unintentional error may be slim.

In his work on self-deception, evolutionary biologist Robert Trivers has investigated the sorts of excuses that scientists make—to themselves and others—for doing questionable work.[5] In an intriguing follow up to his book, Trivers dives into the question of why so many psychologists fail to share data, in contravention of the mandate of the APA-sponsored journals in which they are published.[6] The mere fact that these scientists refuse to share their data is in and of itself shocking. The research that Trivers cites reports that 67 percent of the psychologists who were asked to share their data failed to do so.[7] The intriguing next step came when it was hypothesized that failure to share data might correlate with a higher rate of statistical error in the published papers. When the results were analyzed, it was found not only that there was a higher error rate in those papers where authors had withheld their data sets—when compared to those who had agreed to share them—but that 96 percent of the errors were in the scientists' favor! (It is important to note here that no allegation of fraud was being made against the authors; indeed, without the original data sets how could one possibly check this? Instead the checks were for statistical errors in the reported studies themselves, where Wicherts et al. merely reran the numbers.)

Trivers reports that in another study investigators found further problems in psychological research, such as spurious correlations and running an experiment until a significant result was reached.[8] Here the researchers' attention focused on the issue of "degrees of freedom" in collecting and analyzing data, consonant with problems (3), (4), and (5) above. "Should more data be collected? Should some observations be excluded? Which conditions should be combined and which ones compared? Which control variables should be considered? Should specific measures be combined or transformed or both?" As the researchers note, "[it] is rare, and sometimes impractical, for researchers to make all these decisions beforehand."[9] But this can lead to the "self-serving interpretation of ambiguity" in the evidence. Simmons et al. demonstrate just this by intentionally adjusting the "degrees of freedom" in two parallel studies to "prove" an obviously false result (that listening to a particular song can change the listener's birth date).[10]

I suppose the attitude one might take here is that one is shocked—shocked—to find gambling in Casablanca. Others might contend that psychology is not really a science. But, as Trivers points out, such findings at least invite the question of whether such dubious statistical and methodological techniques can be found in other sciences as well. If psychologists—who study human nature for a living—are not immune to the siren song of self-serving data interpretation, why should others be?

A third type of error in scientific reasoning occurs as a result of the kind of unintentional cognitive biases we all share as human beings, to which scientists are not immune. Preference for one's own theory is perhaps the easiest to understand. But there are literally hundreds of others—representativeness bias, anchoring bias, heuristic bias, and the list goes on—that have been identified and explained in Daniel Kahneman's brilliant book *Thinking Fast and Slow*, and in the work of other researchers in the rapidly growing field of behavioral economics.[11] We will explore the link between such cognitive bias and the paroxysm of science denialism in recent years in chapter 8. That such bias might also have a foothold among those who actually *believe* in science is, of course, troubling.

One might expect that these sorts of biases would be washed out by scientists who are professionally trained to spot errors in empirical reasoning. Add to this the fact that no scientist would wish to be publicly embarrassed by making an error that could be caught by others, and one imagines that the incentive would be great to be objective about one's work and to test one's theories before offering them for public scrutiny. The truth, however, is that it is sometimes difficult for individuals to recognize these sorts of shortcomings in their own reasoning. The cognitive pathways for bias are wired into all of us, PhD or not.

Perhaps the most virulent example of the destructive power of cognitive bias is *confirmation bias*. This occurs when we are biased in favor of finding evidence that confirms what we already believe and discounting evidence that does not. One would think that this kind of mistake would be unlikely for scientists to make, since it goes directly against the scientific attitude, where we are supposed to be searching for evidence that can force us to *change* our beliefs rather than ratify them. Yet again, no matter how much individual scientists might try to inoculate themselves against this type of error, it is sometimes not found until peer review or even post-publication.[12] Thus we see that the scientific attitude is at its most powerful when embraced by the entire scientific community. The scientific attitude is not merely a matter of conscience or reasoning ability among individual scientists, but the cornerstone of those practices that make up science as an institution. As such, it is useful as a guard against all manner of errors, whatever their source.

Critical Communities and the Wisdom of Crowds

Ideally, the scientific attitude would be so well ensconced at the individual level that scientists could mitigate all possible sources of error in their own work. Surely some try to do this, and it is a credit to scientific research that it is one of the few areas of human endeavor where the practitioners are motivated to find and correct their own mistakes by comparing them to actual evidence. But it is too much to think that this happens in every instance. For one thing, this would work only if all errors were unintentional and the researcher was motivated (and able) to find them. But in cases of fraud or willful ignorance, we would be foolish to think that scientists can and will correct all of their own mistakes. For these, membership in a larger community is crucial.

Recent research in behavioral economics has shown that groups are often better than individuals at finding errors in reasoning. Most of these experiments have been done on the subject of unconscious cognitive bias, but it goes without saying that groups also would be better motivated to find errors created by conscious bias than the people who had perpetrated them. Some of the best experimental work on this subject has been done with logic puzzles. One of the most important results is the Wason selection task.[13] In this experiment, subjects were shown four cards lying flat on a table with the instruction that—although they could not touch the cards—each had a number on one side and a letter of the alphabet on the other. Suppose the cards read 4, E, 7, and K. Subjects were then given a rule such as "If there is a vowel on one side of the card, then there is an even number on the other." Their task was to determine which (and only which) cards they would need to turn over in order to test the rule.

The experimenters reported that subjects found this task to be incredibly difficult. Indeed, just 10 percent got the right answer, which is that only the E and the 7 needed to be turned over. The E needs to be turned over because it might have an odd number on the other side, which would disprove the rule. The 7 also needs to be checked, because if there were a vowel on its other side, the rule would also be falsified. Subjects tended to be flummoxed by the fact that the 4 did not need to be turned over. For those who have studied logic, it might be obvious that the 4 is irrelevant to the truth of the rule, because it does not matter what is on the other side of the card. The rule only says that an even number would appear *if* there is a vowel on the other side of the card; it does not say that an even number can appear *only if* there is a vowel on the other side. Even if the other side of the card had a consonant, it would not falsify the rule for a 4 to appear on its obverse. Likewise, one need not check the card with the K. Again, the rule says what must follow *if* there is a vowel on one side of the card; it says nothing about what must be the case if there is not.

The really exciting part of the experiment, however, is what happens when one asks subjects to solve this problem in a group setting. Here 80 percent get the right answer. Note that this is *not* because everyone else in the group merely defers to the "smartest person." The experimenters checked for this and found that even in groups where none of the individual subjects could solve the task, the group often could. What this suggests is that when we are placed in a group, our reasoning skills improve. We scrutinize one another's hypotheses and criticize one another's logic. By mixing it up we weed out the wrong answers. In a group we are open to persuasion, but also to persuading others. As individuals, we may lack the motivation to criticize our own arguments. Once we have arrived at an answer that "seems right," why would we bother to rethink it? In groups, however, there is much more scrutiny and, as a result, we are much more likely to arrive at the right answer. This is not because someone at the back of the room knows it, but because human reasoning skills seem to improve when we are placed within a group.[14]

In his book *Infotopia: How Many Minds Produce Knowledge*,[15] Cass Sunstein explores the idea—based on empirical evidence such as that cited above—that there are myriad benefits to collective reasoning. This is sometimes referred to as the "wisdom of crowds" effect, but this populist moniker does not do justice to the case of science, where we correctly value expert opinion. As we shall see, scientific practice is nonetheless well described by Sunstein, who acknowledges the power of experts. And there are other effects as well. In his book, Sunstein explores three principle findings, each backed up by experimental evidence. On the whole, in human reasoning:

(1) groups do better than individuals,

(2) interactive groups do better than aggregative groups, and

(3) experts do better than laypeople.[16]

As we saw with the Wason selection task, the findings in numbers (1) and (2) are on full display. Groups do better than individuals not only because someone in the room might have the right answer (which would presumably be recognized once it was arrived at) but because even in those situations in which no one knows the right answer, the group can interact and ask critical questions, so that they may discover something that none of them individually knew. In this case, the collective outcome can be greater than what would have been reached by any of its individual members (or a mere aggregation of group opinion).

There is of course some nuance to this argument, and one must state the limiting conditions, for there are some circumstances that undermine the general result. For instance, in groups where there is heavy pressure to defer to authority, or there is strict hierarchy, groups can succumb to "group think." In such cases, the benefits of group reasoning can be lost to the "cascade effect" of whomever speaks first, or to an "authority effect" where one subordinates one's opinion to the highest ranking member of the group. Indeed, in such cases the aggregative effect of groups can even be a drawback, where everyone expresses the same opinion—whether they believe it or not—and group reasoning converges on falsehood rather than truth. To remedy this, Sunstein proposes a set of principles to keep groups at their most productive. Among them:

(1) groups should regard dissent as an obligation,

(2) critical thinking should be prized, and

(3) devil's advocates should be encouraged.[17]

This sounds a good deal like science. Especially in a public setting, scientists are notably competitive with one another. Everyone wants to be right.

There is little deference to authority. In fact, the highest praise is often reserved not for reaching consensus but for finding a flaw in someone else's reasoning. In such a super-charged interactive environment, Sunstein's three earlier findings may reasonably be combined into one where we would expect that *groups* of *experts* who *interact* with one another would be the most likely to find the right answer to a factual question. Again, this sounds a lot like science.

I think it is important, however, not to put too much emphasis on the idea that it is the *group* aspect of science that makes so much difference as the fact that science is an *open* and *public* process. While it is true that scientists often work in teams and participate in large conferences and meetings, even the lone scientist critiquing another's paper in the privacy of her office is participating in a community process. It is also important to note that even though these days it is more common for scientists to collaborate on big projects like the Large Hadron Collider, this could hardly be the distinguishing feature of what makes science special, for if it were then what to say about the value of those scientific theories put forth in the past by sole practitioners like Newton and Einstein? Working in groups is one way of exposing scientific ideas to public scrutiny. But there are others. Even when working alone, one knows that before a scientific theory can be accepted it still must be vetted by the larger scientific community. *This* is what makes science distinctive: the fact that its values are embraced by a wider community, not necessarily that the practice of science is done in groups.

It is the hallmark of science that individual ideas—even those put forth by experts—are subjected to the highest level of scrutiny by other experts, in order to discover and correct any error or bias, whether intentional or not. If there is a mistake in one's procedure or reasoning, or variance with the evidence, one can expect that other scientists will be motivated to find it. The scientific attitude is embraced by the community as embodied in a set of practices. It is of course heartening to learn from Sunstein and others that experimental work in psychology has vindicated a method of inquiry that is so obviously consonant with the values and practices of science. But the importance of these ideas has long been recognized by philosophers of science as well.

In a brilliant paper entitled "The Rationality of Science versus the Rationality of Magic,"[18] Tom Settle examines what it means to say that scientific belief is rational whereas belief in magic is not. The conclusion he comes to is that one need not disparage the individual rationality of members of those communities that believe in magic any more than one should attempt to explain the distinctiveness of science through the rationality of individual scientists. Instead, he finds the difference between science and magic in the "corporate critical-mindedness" of science, which is to say that there is a tradition of *group* criticism of individual ideas that is lacking in magic.[19] As Hansson explains in his article "Science and Pseudo-Science,"

> [According to Settle] it is the rationality and critical attitude built into institutions, rather than the personal intellectual traits of individuals, that distinguishes science from non-scientific practices such as magic. The individual practitioner of magic in a pre-literate society is not necessarily less rational than the individual scientists in modern Western society. What she lacks is an intellectual environment of collective rationality and mutual criticism.[20]

As Settle explains further:

> I want to stress the institutional role of criticism in the scientific tradition. It is too strong a requirement that every individual within the scientific community should be a first rate critic, especially of his own ideas. In science, criticism may be predominantly a communal affair.[21]

Noretta Koertge too finds much to praise in the "critical communities" that are a part of science. In her article "Belief Buddies versus Critical Communities," she writes:

> I have argued that one characteristic differentiating typical science from typical pseudoscience is the presence of critical communities, institutions that foster communication and criticism through conferences, journals, and peer review.... We have a romantic image of the lone scientist working in isolation and, after many years, producing a system that overturns previous misconceptions. We forget that even the most reclusive of scientists these days is surrounded by peer-reviewed journals; and if our would-be genius does make a seemingly brilliant discovery, it is not enough to call a news conference or promote it on the web. Rather, it must survive the scrutiny and proposed amendments of the relevant critical scientific community.[22]

Finally, in the work of Helen Longino, we see respect for the idea that, even if one views science as an irreducibly social enterprise—where the values of its individual practitioners cannot help but color their scientific work—it is the collective nature of scientific practice as a whole that helps to support its objectivity. In her pathbreaking book *Science as Social Knowledge*,[23] Longino embraces a perspective that one may initially think of as hostile to the claim that scientific reasoning is privileged. Her overall theme seems consonant with the social constructivist response to Kuhn: that scientific inquiry, like all human endeavors, is value laden, and therefore cannot strictly speaking be objective; it is a myth that we choose our beliefs and theories based only on the evidence.

Indeed, in the preface to her book, Longino writes that her initial plan had been to write "a philosophical critique of the idea of value-free science." She goes on, however, to explain that her account grew to be one that "reconciles the objectivity of science with its social and cultural construction." One understands that her defense will not draw on the standard empiricist distinction between "facts and values." Nor is Longino willing to situate the defense of science in the logic of its method. Instead, she argues for two important shifts in our thinking about science: (1) that we should embrace the idea of seeing science as a practice, and (2) that we should recognize science as practiced not primarily by individuals but by social groups.[24] But once we have made these shifts a remarkable insight results, for she is able to conclude from this that "the objectivity of scientific inquiry is a consequence of this inquiry's being a social, and not an individual, enterprise."

How does this occur? Primarily through recognition that the diversity of interests held by each of the individual scientific practitioners may grow into a system of checks and balances whereby peer review before publication, and other scrutiny of individual ideas, acts as a tonic to individual bias. Thus scientific knowledge becomes a "public possession."

> What is called scientific knowledge, then, is produced by a community (ultimately the community of all scientific practitioners) and transcends the contributions of any individual or even of any subcommunity within the larger community. Once propositions, theses, and hypotheses are developed, what will become scientific knowledge is produced collectively through the clashing and meshing of a variety of points of view....
>
> Objectivity, then, is a characteristic of a community's practice of science rather than of an Individual's, and the practice of science is understood in a much broader sense than most discussions of the logic of scientific method suggest.[25]

She concludes, "values are not incompatible with objectivity, but objectivity is analyzed as a function of community practices rather than as an attitude of individual researchers toward their material."[26] One can imagine no better framing of the idea that the scientific attitude must be embraced not just by individual researchers but by the entire scientific community.

It should be clear by now how the conclusions of Settle, Koertge, and Longino tie in neatly with Sunstein's work. It is not just the honesty or "good faith" of the individual scientist, but fidelity to the scientific attitude as community practice that makes science special as an institution. No matter the biases, beliefs, or petty agendas that may be put forward by individual scientists, science is more objective than the sum of its individual practitioners. As philosopher Kevin deLaplante states, "Science as a social institution is distinctive in its commitment to reducing biases that lead to error."[27] But how precisely does science do this? It is time now to take a closer look at some of the institutional techniques that scientists have developed to keep one another honest.

Methods of Implementing the Scientific Attitude to Mitigate Error

It would be easy to get hung up on the question of the sources of scientific error. When one finds a flaw in scientific work—whether it was intentional or otherwise—it is natural to want to assign blame and examine motives. Why are some studies irreproducible? Surely not all of them can be linked to deception and fraud. As we have seen, there also exists the possibility of unconscious cognitive bias that is a danger to science. It is wrong to think of every case in which a study is flawed as one of corruption.[28] Still, the problem has to be fixed; the error needs to be caught. The important point at present is not where the error comes from but that science has a way to deal with it.

We have already seen that groups are better than individuals at finding scientific errors. Even when motivated to do so, an individual cannot normally compete with the "wisdom of crowds" effect among experts. Fortunately, science is set up to bring precisely this sort of group scrutiny to scientific hypotheses. Science has institutionalized a plethora of techniques to do this. Three of them are *quantitative methods*, *peer review*, and *data sharing and replication*.

Quantitative Methods

There are entire books on good scientific technique. It is important to point out that these exist not just for quantitative research, but for qualitative investigation as well.[29] There are some time-honored tropes in statistical reasoning that one trusts any scientist would know. Some of these—such as that there is a difference between causation and correlation—are drilled into students in Statistics 101. Others—for instance, that one should use a different data set for creating a hypothesis than for testing it—are a little more subtle. With all of the rigor involved in this kind of reasoning, there is little excuse for scientists to make a quantitative error. And yet they do. As we have seen, the tendency to withhold data sets is associated with mathematical errors in published work. Given that everyone is searching for the holy grail of a 95 percent confidence level that indicates statistical significance, there is also some fudging.[30] But publicity over any malfeasance—whether it is due to fraud or mere sloppiness—is one of the ways that the scientific attitude is demonstrated. Scientists check one another's numbers. They do not wait to find an error; they go out and look for one. If it is missed at peer review, some errors are found within hours of publication. I suppose one could take it as a black mark against science that *any* quantitative or analytical errors ever occur. But it is better, I think, to count it as a virtue of science that it has a culture of finding such errors—where one does not just trust in authority and assume that everything is as stated. Yet there are some methodological errors that are so endemic—and so insidious—that they are just beginning to come to light.[31]

Statistics provides many different tests one can do for evidential relationships. Of course, none of these will ever amount to causation, but correlation is the coin of the realm in statistics, for where we find correlation we can sometimes infer causation. One of the most popular calculations is to determine the *p-value* of a hypothesis, which is the probability of finding a particular result if the *null hypothesis* were true (which is to say if there were no real-world correlation between the variables). The null hypothesis is the working assumption against which statistical significance is measured. It is the "devil's advocate" hypothesis that no actual causal relationship exists between two variables, which means that if one finds any such correlation it will be the result of random chance. To find statistical significance, one must therefore find strong enough statistical evidence to reject the null hypothesis—one must show that the correlation found is greater than what one would expect from chance alone. By convention, scientists have chosen 0.05 as the inflection point to indicate statistical significance, or the likelihood that a correlation would not have occurred just by chance. When we have reached this threshold, it suggests a higher likelihood of real-world correlation.[32] The p-value is therefore the probability that you would get your given data if the null hypothesis were true. A small p-value indicates that it is highly improbable that a correlation is due to chance; the null hypothesis is probably wrong. This is what scientists are looking for. A large p-value is just the opposite, indicating weak evidence; the null hypothesis is probably true. A p-value under 0.05 is therefore highly sought after, as it is customarily the threshold for scientific publication.

P-hacking (also known as data dredging) is when researchers gather a large amount of data, then sift through it looking for anything that might be positively correlated.[33] As understood, since there is a nonzero probability that two things may be correlated by chance, this means that if one's data set is large enough, and one has enough computing power, one will almost certainly be able to find some positive correlation that meets the 0.05 threshold, whether it is a reflection of a real-world connection or not. This problem is exacerbated by the degrees of freedom we talked about earlier, whereby researchers make decisions about when to stop collecting data, which data to exclude, whether to hold a study open to seek more data, and so on. If one looks at the results midway through a study to decide whether to continue collecting data, that is p-hacking.[34] Self-serving exploitation of such degrees of freedom can be used to manipulate almost any data set into some sort of positive correlation. And these days it is much easier than it used to be.[35] Now one merely has to run a program and the significant results pop out. This in and of itself raises the likelihood that some discovered results will be spurious. We don't even need to have a prior hypothesis that two things are related anymore. All we need are raw data and a fast computer. An additional problem is created when some researchers choose to selectively report their results, excluding all of the experiments that fall below statistical significance. If one treats such decisions about what to report and what to leave out as one's final degree of freedom, trouble can arise. As Simmons et al. put it in their original paper: "it is unacceptably easy to publish 'statistically significant' evidence consistent with *any* hypothesis."[36]

Examples surround us. Irreproducible studies on morning TV shows that report correlations between eating grapefruit and having an innie belly button, or between coffee and cancer, rightly make us suspicious. The definitive example, however, was given in Simmons's original paper, where researchers were able—by judicious manipulation of their degrees of freedom—to prove that listening to the Beatles song "When I'm 64" had an effect on the listener's age. Note that researchers did not show that listening to the song made one *feel* younger; they showed that it actually *made* one younger.[37] Demonstrating something that is causally impossible is the ultimate indictment of one's analysis. Yet p-hacking is not normally considered to be fraud.

> In most cases researchers could be honestly making necessary choices about data collection and analysis, and they could really believe they are making the correct choices, or at least reasonable choices. But their bias will influence those choices in ways that researchers may not be aware of. Further, researchers may simply be using the techniques that "work"—meaning they give the results the researcher wants.[38]

Add to this the pressures of career advancement, competition for research funding, and the "publish or perish" environment to win tenure or other career advancement at most colleges and universities, and one has created an ideal motivational environment to "make the correct choices" about one's work. As Uri Simonsohn, one of the coauthors of the original study put it: "everyone has p-hacked a little."[39] Another commentator on this debate, however, sounded a more ominous warning in the title of his paper: "Why Most Published Research Findings Are False."[40]

But is all of this really so bad? Of course scientists only want to report positive results. Who would want to read through all of those failed hypotheses or see all the data about things that didn't work? Yet this kind of selective reporting—some might call it "burying"—is precisely what some pseudoscientists do to make themselves appear more scientific. In Ron Giere's excellent book *Understanding Scientific Reasoning* he outlines one such example, whereby Jeane Dixon and other "futurists" often brag of being able to predict the future, which they do by making thousands of predictions about the coming year, then reporting only the ones that came true at year's end.[41]

To many, all of this will seem to be a bastardization of the scientific attitude because, although one is looking at the evidence, it is being used in an inappropriate way. With no prior hypothesis in mind to test, evidence is gathered merely for the sake of "data" to be mined. If one is simply on a treasure hunt for "p"–for something that looks significant even though it probably isn't—where is the warrant for one's beliefs? One can't help but think here of Popper and his admonition about positive instances. If one is looking for positive instances, they are easy to find! Furthermore, without a hypothesis to test, one has done further damage to the spirit of scientific investigation. We are no longer building a theory—and possibly changing it depending on what the evidence says—but only reporting correlations, even if we don't know why they occur or may even suspect that they are spurious.

Clearly, p-hacking is a challenge to the thesis that scientists embrace the scientific attitude. But unless one is willing to discard a large proportion of unreproducible studies as unscientific, what other choice do we have? Yet remember my claim that what makes science special is not that it never makes mistakes, but that it responds to them in a rigorous way. And here the scientific community's response to the p-hacking crisis strongly supports our confidence in the scientific attitude.

First, it is important to understand the scope of p-value, and the job that science has asked of it. It was never intended to be the sole measure of significance.

> [When] Ronald Fisher introduced the P value in the 1920s, he did not mean it to be a definitive test. He intended it simply as an informal way to judge whether evidence was significant in the old-fashioned sense: worthy of a second look. The idea was to run an experiment, then see if the results were consistent with what random chance might produce.... For all the P value's apparent precision, Fisher intended it to be just one part of a fluid, non-numerical process that blended data and background knowledge to lead to scientific conclusions. But it soon got swept into a movement to make evidence-based decision-making as rigorous and objective as possible.... "The P value was never meant to be used the way it's used today."[42]

Another problem with the p-value is that it is commonly misunderstood, even by the people who routinely use it:

> The p-value is easily misinterpreted. For example, it is often equated with the strength of a relationship, but a tiny effect size can have very low p-values with a large enough sample size. Similarly, a low p-value does not mean that a finding is of major ... interest.[43]

In other words, p-values cannot measure the size of an effect, only the probability that the effect occurred by chance. Other misconceptions surround how to interpret the probabilities expressed by a p-value. Does a study with a 0.01 p-value mean that there is just a 1 percent chance of the result being spurious?[44] Actually, no. This cannot be determined without prior knowledge of the likelihood of the effect. In fact, according to one calculation, "a P value of 0.01 corresponds to a false-alarm probability of at least 11%, depending on the underlying probability that there is a true effect."[45]

All this may lead some to wonder whether the p-value is as important as some have said. Should it be the gold standard of statistical significance and publication? Now that the problems with p-hacking are beginning to get more publicity, some journals have stopped asking for it.[46] Other critics have proposed various statistical tests to detect p-hacking and shine a light on it. Megan Head et al. have proposed a method of *text mining* to measure the extent of p-hacking. The basic idea here is that if someone is p-hacking, then the p-values in their studies will cluster around the 0.05 level, which will result in an odd shape if one curves the results.[47]

The next step might be to change the reporting requirements in journals, so that authors are required to compute their own p-curves, which would enable other scientists to tell at a glance whether the results had been p-hacked. Other ideas have included mandatory disclosure of all degrees of freedom taken in producing the result of a paper, along with size of the effect and any information about prior probabilities.[48] Some of this is controversial because researchers cannot agree on the usefulness of what is reported.[49] And, as Head points out:

> Many researchers have advocated abolishing NHST [Null Hypothesis Significance Testing]. However, others note that many of the problems with publication bias reoccur with other approaches, such as reporting effect sizes and their confidence intervals or Bayesian credible intervals. Publication biases are not a problem with p-values per se. They simply reflect the incentives to report strong (i.e. significant) facts.[50]

So perhaps the solution is full disclosure and transparency? Maybe researchers should be able to use p-values, but the scientific community who assesses their work—peer reviewers and journal editors—should be on the lookout. In their paper, Simmons et al. propose several guidelines:

Requirements for authors:

1. Authors must decide the rule for terminating data collection before data collection begins and report this rule in the article.
2. Authors must collect at least 20 observations per cell or else provide a compelling cost-of-data collection justification.
3. Authors must list all variables collected in a study.
4. Authors must report all experimental conditions, including failed manipulations.
5. If observations are eliminated, authors must also report what the statistical results are if those observations are included.
6. If an analysis includes a covariate, authors must report the statistical reports of the analysis without the covariate.

Guidelines for reviewers:

1. Reviewers should ensure that authors follow the requirements.
2. Reviewers should be more tolerant of imperfection in results.
3. Reviewers should require authors to demonstrate that their results do not hinge on arbitrary analytic decisions.
4. If justifications of data collection or analysis are not compelling, reviewers should require the authors to conduct an exact replication.[51]

And then, in a remarkable demonstration of the efficacy of these rules, Simmons et al. went back to their bogus hypothesis—about the Beatles song that "changed" the listener's age—and redid it according to the above guidelines ... and the effect disappeared. Can one imagine any pseudoscience taking such a rigorous approach to ferreting out error?

The amazing thing about p-hacking is that even though it reflects a current crisis in science, it still makes a case for the value of the scientific attitude. It would have been easy for me to choose some less controversial or embarrassing example to demonstrate that the community of scientists have embraced the scientific attitude through their insistence on better quantitative methods. But I didn't. I went for the worst example I could find. Yet it still came out that scientists were paying close attention to the problem and trying to fix it. Even though one of the procedures of science is flawed, the response from the scientific community has been pitch perfect: *We've got this. We may not have it solved yet, and we may make more mistakes, but we are investigating the problem and trying to correct it. Although it may be in an individual scientist's interests to be sloppy or lazy (or worse) and rely on p-hacking, this is an embarrassment to the scientific community at large. This is not who we are and we will correct this.*

At the end of his paper, Simmons says this:

> Our goal as scientists is not to publish as many articles as we can, but to discover and disseminate truth. Many of us—and this includes the three authors of this article—often lose sight of this goal, yielding to the pressure to do whatever is justifiable to compile a set of studies that we can publish. This is not driven by a willingness to deceive but by the self-serving interpretation of ambiguity, which enables us to convince ourselves that whichever decisions produced the most publishable outcome must have also been the most appropriate. This article advocates a set of disclosure requirements that imposes minimal costs on authors, readers, and reviewers. These solutions will not rid researchers of publication pressures, but they will limit what authors are able to justify as acceptable to others and to themselves. We should embrace these disclosure requirements as if the credibility of our profession depended on them. Because it does.[52]

This warning is not lost on others. At the end of his own article "P-Hacking and Other Statistical Sins," Steven Novella lists a set of general principles for scientific conduct, then brings up an odious comparison that should make all philosophers of science sit up and take notice:

> Homeopathy, acupuncture, and ESP research are all plagued by ... deficiencies. They have not produced research results anywhere near the threshold of acceptance. Their studies reek of P-hacking, generally have tiny effect sizes, and there is

> no consistent pattern of replication.... But there is no clean dichotomy between science and pseudoscience. Yes, there are those claims that are far toward the pseudoscience end of the spectrum. All of these problems, however, plague mainstream science as well. The problems are fairly clear, as are the necessary fixes. All that is needed is widespread understanding and the will to change entrenched culture.[53]

One can imagine no better testament—or call to arms—for the scientific attitude than that it aspires to create such a culture.[54]

Peer Review

We have already seen some of the advantages of peer review in the previous section. Where individuals may be tempted to take shortcuts in their work, they will be caught up short by representatives of the scientific community: the expert readers who agree to provide an opinion and comments on an anonymous manuscript, receiving for compensation only the knowledge that they are helping the profession and that their identity will be shielded.[55] (There is also the advantage of getting to see new work first and having the chance to influence it through one's comments.) But that is it. It is customarily shocking to outsiders to realize that most scientific review work is done for free, or at most for minimal compensation such as a free journal subscription or book.

The process of peer review is *de riguer* at most scientific journals, where editors would shudder to share work that had not been properly vetted. In most instances, by the way, this means that every journal article is reviewed by not one but *two* experts, to provide an extra level of scrutiny. A check on a check. To the consternation of most authors, publication normally requires that both referees give the nod. And if any changes are proposed, it is common practice to send the paper back to the original reviewers once changes are made. Inferior work does sometimes slip through, but the point is that this system is one in which the scientific attitude is demonstrated through the principles of fair play and objectivity. It is not perfect, but it is at least a mechanism that allows the scientific community as a whole to have a hand in influencing individual research results before they are shared with the entire community.

And there is also another level of scrutiny, for if a mistake is made there is always *retraction*. If an error is not caught until after a journal article is published, the mistake can be flagged or the whole article can be taken back, with due publicity for future readers. Indeed, in 2010 two researchers, Ivan Oransky and Adam Marcus, founded a website called RetractionWatch.com where one can find an up-to-date list of scientific papers that have been retracted. The website was created because of concerns that retracted papers were not being given enough publicity, which might lead other scientists mistakenly to build upon them. Again, one can bemoan the fact that approximately six hundred scientific papers per year are retracted, or one can applaud the scientific community for making sure that the problem does not become worse. By publicizing retractions, it might even provide an incentive for researchers not to end up on such a public "wall of shame." On their blog, Oransky and Marcus argue that Retraction Watch contributes in part to the self-correcting nature of science.

Although many see retraction as an embarrassment (and it surely is for the parties involved), it is yet another example of the scientific attitude in action. It is not that mistakes never occur in science, but when they do there are well-publicized community standards by which these errors can be rectified. But it is important to note, although it is sometimes lost on the larger nonscientific community, that article retraction does not necessarily indicate fraud. It is enough merely for there to be an error large enough to undermine the original finding. (In some cases this is discovered when other scientists try to reproduce the work, which will be discussed in the next section.) The example of p-hacking alone surely indicates that there is enough to keep the watchdogs busy. But fraud is the ultimate betrayal of the scientific attitude (and as such it will be discussed in its own chapter). For now, it is enough to note that article retraction can occur for many reasons, all of which credit the vigilance of the scientific community.

Peer review is thus an essential part of the practice of science. There is only so far that individual scientists can go in catching their own errors. Maybe they will miss some. Maybe they want to miss some. Having someone else review work before publication is the best possible way to catch errors and keep them from infecting the work of others. Peer review keeps one honest, even if one was honest in the first place. Remember that groups outperform individuals. And groups of experts outperform individual experts. Even before these principles were experimentally proven, they were incorporated into the standard of peer review. What happens when an idea or theory does not go through peer review? Maybe nothing. Then again, on the theme that perhaps we can learn most about science by looking at its failures, I propose that we here consider the details of an example I introduced in chapter 3, which demonstrates the pitfalls of forgoing community scrutiny before publication: cold fusion.

To those who know only the "headline" version of cold fusion, it may be tempting to dismiss the whole episode as one of fraud or at least bad intent. With the previously noted book titles *Bad Science* and *Cold Fusion: The Scientific Fiasco of the Century* it is easy to understand why people would want to cast aspersions on the two original researchers, then try to put the whole thing on a shelf and claim it could never happen again. What is fascinating, however, is the extent to which this episode illuminates the process of science. Whatever the ultimate conclusion after it was over—and there was plenty of armchair judgment after upwards of $50 million was spent to refute the original results and a governmental panel was appointed to investigate[56]—the entire cold fusion fiasco provides a cautionary tale for what can happen when the customary practices of science are circumvented.

The problems in this case are so thick that one might use the example of cold fusion to illustrate any number of issues about science: weakness in analytical methods, the importance of data sharing and reproducibility, the challenges of subjectivity and cognitive bias, the role of empirical evidence, and the dangers of interference by politics and the media. Indeed, one of the books written on cold fusion has a final chapter that details the numerous lessons we can learn about the scientific process based on the failures demonstrated by cold fusion.[57] Here I will focus on only one of these: peer review.

As noted, on March 23, 1989, two chemists, B. Stanley Pons and Martin Fleischmann, held a press conference at the University of Utah to announce a startling scientific discovery: they had experimental evidence for the possibility of conducting a room temperature nuclear fusion reaction, using materials that were available in most freshman chemistry labs. The announcement was shocking. If true, it raised the possibility that commercial adaptations could be found to produce a virtually unlimited supply of energy. Political leaders and journalists were mesmerized, and the announcement got front-page coverage in several media outlets, including the *Wall Street Journal*. But the press conference was shocking for another reason, which was that Pons and Fleischmann had as yet produced no scientific paper detailing their experiment, thus forestalling the possibility of other scientists checking their work.

To say that this is unusual in science is a vast understatement. Peer review is a bedrock principle of science, because it is the best possible way of catching any errors—and asking all of the appropriate critical questions—before a finding is released. One commentator in this debate, John Huizenga, acknowledges that "the rush to publish goes back to the seventeenth century," when the Royal Society of London began to give priority to the first to publish, rather than the first to make a discovery.[58] The competitive pressures in science are great, and priority disputes for important findings are common. Thus, "in the rush to publish, scientists eliminate important experimental checks of their work and editors compromise the peer-review process, leading to incomplete or even incorrect results."[59] And this seems to be what happened with cold fusion.

At the time of Pons and Fleischmann's press conference, they were racing to beat another researcher from a nearby university, whom they claim had pirated their discovery when he had read about it during review of their grant application. That researcher, Steven Jones of Brigham Young University, counter-claimed that he had been working on fusion research long before this, and that Pons and Fleischmann's grant proposal was merely an "impetus" to get back to work and brook no further delay in publishing his own results. Once the administration at their respective universities (BYU and Utah) became involved, and the subject turned to patents and glory for the institution, things devolved and few behaved admirably. As noted, scientists are human beings and feel the same competitive pressures that any of us would. But Pons and Fleischmann would live to regret their haste.

As it turned out, their experimental evidence was weak. They claimed to have turned up an enormous amount of excess heat by conducting an electrochemical experiment involving palladium, platinum, a mixture of heavy water and lithium, and some electric current. But, if this was a nuclear reaction and not a chemical one, they had a problem: where was the radiation? In the fusion of two deuterium nuclei, it is expected that not only heat but also neutrons (and gamma rays) will be given off. Yet repeatedly, Pons and Fleischmann failed to find any neutrons (or so few as to suggest that there must be some sort of error) with the detector they had borrowed from the physics department. Could there be some other explanation?

One would think that it would be a fairly straightforward process to sort all of this out, but in the earliest days after the press conference, Pons and Fleischmann refused to share their data with other experimenters, who were reduced to "[getting their] experimental details from *The Wall Street Journal* and other news publications."[60] Needless to say, that is not how science is supposed to work. Replications are not supposed to be based on guesswork about experimental technique or phone calls begging for more information. Nonetheless, in what Gary Taubes calls a "collective derangement of minds," a number of people in the scientific community got caught up in the hype, and several partial "confirmations" began to pop up around the world.[61] Several research groups reported that they too had found excess heat in the palladium experiment (but still no radiation). At the Georgia Tech Research Institute, they claimed to have found an explanation for the absence of deuterium atoms due to the presence of Boron as an ingredient in the Pyrex glassware that Pons and Fleischmann had used; when they tested the result with Boron-free glassware they detected some radiation, which disappeared when the detector was screened behind a Boron shield. Another group claimed to have a theoretical explanation for the lack of radiation, with helium and heat as the only product of Pons and Fleischmann's electrolysis reaction.[62]

Of course, there were also critics. Some nuclear physicists in particular were deeply skeptical of how a process like fusion—which occurs at enormous temperature and pressure in the center of the Sun—could happen at room temperature. Many dismissed these criticisms, however, as representative of the "old" way of thinking about fusion. A psychologist would have a field day with all of the tribalism, bandwagon jumping, and "emperor has no clothes" responses that happened next. Indeed, some have argued that a good deal of the positive response to cold fusion can be explained by Solomon Asch's experimental work on conformity, in which he found that a nontrivial number of subjects would deny self-evident facts if that was necessary to put them in agreement with their reference group.[63] It is important to point out, however, that although all of these psychological effects may occur in science, they are supposed to be washed out by the sort of critical scrutiny that is performed by the community of scientists. So where was this being done? While it is easy to cast aspersions on how slowly things developed, it is important to note that the critical research *was* being done elsewhere, as other researchers worked under the handicap of inadequate data and too many unanswered questions. And, as some details about Pons and Fleishmann's work began to leak out, the problems for cold fusion began to mount.

For one, where was the radiation? As noted, some thought they could explain its absence, yet most remained skeptical. Another problem arose as some questioned whether the excess heat could be explained as a mere artifact of some unknown chemical process. On April 10, 1989, a rushed and woefully incomplete paper by Pons and Flesichmann appeared in the *Journal of Electroanalytical Chemistry*. It was riddled with errors. Why weren't these caught at peer review? Because the paper was not peer reviewed![64] It was evaluated only by the editor of the journal (who was a friend of Stanley Pons) and rushed into print, because of the "great interest" of the scientific community.[65] At numerous scientific conferences to follow, Pons was reluctant to answer detailed questions from colleagues about problems with his work and refused to share raw data. In his defense, one might point out that by this point numerous partial "confirmations" had been coming in and others were defending his work too, claiming that they had similar results. Even though Pons was well aware of some of the shortcomings of his work, he might still have believed in it.

And yet, as we saw earlier in this chapter, there is a high correlation between failure to share one's data and analytical error.[66] It is unknown whether Pons deliberately tried to keep others from discovering something that he already knew to be a weakness or was merely caught up in the hype surrounding his own theory. Although some have called this fraud—and it may well be—one need not go this far to find an indictment of his behavior. Scientists are expected to participate—or at least cooperate—in the criticism of their own theory, and woe to those who stonewall. Pons is guilty at least of obstructing the work of others, which is an assault against the spirit of the scientific attitude, where we should be willing to compare our ideas against the data, whether that is provided by oneself or others.[67]

The process did not take long to unravel from there. A longer, more carefully written version of Pons and Fleischmann's experiment was submitted to the journal *Nature*. While waiting for the paper to be peer reviewed, the journal editor John Maddox published an editorial in which he said:

> It is the rare piece of research indeed that both flies in the face of accepted wisdom and is so compellingly correct that its significance is instantly recognized. Authors with unique or bizarre approaches to problems may feel that they have no true peers who can evaluate their work, and feel more conventional reviewers' mouths are already shaping the word "no" before they give the paper much attention. But it must also be said that most unbelievable claims turn out to be just that, and reviewers can be forgiven for perceiving this quickly.[68]

A few weeks later, on April 20, Maddox announced that he would not be publishing Pons and Fleischmann's paper because all three of its peer reviewers had found serious problems and criticisms of their work. Most serious of these was that Pons and Fleishmann had apparently not done any controls! This in fact was not true. In addition to doing the fusion experiment with heavy water, Pons and Fleischmann had also done an early version of the experiment with light water, and had gotten a similar result.[69] Pons kept this fact quiet, however, sharing it only with Chuck Martin of Texas A&M, who had gotten the same result. When they talked, Pons told him that he was now at the "most exciting" part of the research, but he couldn't discuss it further because of national defense issues.[70] Similarly, other researchers who had "confirmed" the original result were finding that it also worked with carbon, tungsten, and gold.[71] But if the controls all "worked," didn't this mean that the original result was suspicious? Or could it mean that Pons and Fleischmann's original finding was even broader—and more exciting—than they had thought?

It didn't take long for the roof to collapse. On May 1—at a meeting of the American Physical Society in Baltimore—Nate Lewis (an electrochemist from Caltech) gave a blistering talk in which he all but accused Pons and Fleischmann of fraud for their shoddy experimental results; he received a standing ovation from the two thousand physicists in attendance. On May 18, 1989 (less than two months after the original Utah press conference), Richard Petrasso et al. (from MIT) published a paper in *Nature* showing that Pons and Fleischmann's original results were experimental artifacts and not due to any nuclear fusion reaction. It seems that the full gamma ray spectrum graph that Pons had kept so quiet (even though it purported to be their main piece of evidence) had been misinterpreted. From there the story builds momentum with even more criticism of the original result, criticism of the "confirmation" of their findings, and a US governmental panel that was assigned to investigate. Eventually Pons disappeared and later resigned.

Peer review is one of the most important ways that the scientific community at large exercises oversight of the mistakes and sloppiness of individuals who may be motivated—either consciously or unconsciously—by the sorts of pressures that could cause them to cut corners in their work. Even when it is handicapped or stonewalled, this critical scrutiny is ultimately unstoppable. Science is a public venture and we must be prepared to present our evidence or face the consequences. In any case, the facts will eventually come out. Some may, of course, bemoan the fact that the whole cold fusion debacle happened *at all* and see it as an indictment of the process of science. Many critics of science will surely be prepared to do so, with perhaps nothing better than pseudoscience to offer in its place. But I think it is important to remind ourselves that—as we saw with p-hacking—the distinctiveness of science does not lie in the claim that it is perfect. Indeed, error and its discovery are an important part of what moves science forward. More than one commentator in this debate has pointed out that science is self-correcting. When an error occurs, it does not customarily fester. If Pons and Fleischmann had been prepared to share their data and embrace the scientific attitude, the whole episode might have been over within a few days, if it ever got started at all. Yet even when egos, money, and other pressures were at stake, the competitive and critical nature of scientific investigation all but guaranteed that any error would eventually be discovered. So again, rather than merely blaming science that an error happened, perhaps we should celebrate the fact that one of the largest scientific errors of the twentieth century was caught and run to ground less than two months after it was announced, based solely on lack of fit with the empirical evidence. For all the extraneous factors that complicated matters and slowed things down, this was the scientific attitude at its finest. As John Huizenga puts it:

> The whole cold fusion fiasco serves to illustrate how the scientific process works. … Scientists are real people and errors and mistakes do occur in science. These are usually detected either in early discussions of ones research with colleagues or in the peer-review process. If mistakes escape notice prior to publication, the published work will come under close scrutiny by other scientists, especially if it disagrees with an established body of data. … The scientific process is self-corrective.[72]

Data Sharing and Replication

As we have just seen, failure to share data is bad. When we sign on as a scientist, we are implicitly agreeing to a level of intellectual honesty whereby we will cooperate with others who are seeking to check our results for accuracy. In some cases (such as those journals sponsored by the American Psychological Association), the publishing agreement actually *requires* one to sign a pledge to share data with other researchers who request it. Beyond peer review, this introduces the standard of expecting that scientific findings should be replicable.

As Trivers indicates in his review of Wicherts's findings, this does not mean that data sharing necessarily occurs. As human beings, we sometimes shirk what is expected of us. As a community, however, this does not change expectations.[73] The scientific attitude requires us to respect the refutatory power of evidence; part of this is being willing to cooperate with others who seek to scrutinize or even refute our hard-won theories. This is why it is so important for this standard to be enforced by the scientific community (and to be publicized when it is not).

It should be noted, though, that there is a distinction between refusing to share data and doing irreproducible work. It is probably expected that one of the motivations for refusing to share data is fear that someone else will not be able to confirm our results. While this does not necessarily mean that our work is fraudulent, it is certainly an embarrassment. It indicates that *something* must be wrong. As we've seen, that might fall into the camp of quantitative error, faulty analysis, faulty method, bad data collection, or a host of other missteps that could be described as sloppy or lazy. Or it might reveal fraud. In any case, none of this is good news for the individual researcher, so some of them hold back their data. Trivers rightly excoriates this practice, saying that such researchers are "a parody of academics." And, as Wicherts demonstrates, there is a troubling correlation between refusing to share data and a higher likelihood of quantitative error in a published study.[74] One imagines that, if the actual data were available, other problems might pop out as well.

Sometimes they do. The first cut is when a second researcher uses the same data and follows the same methods as the original researcher, but the results do not hold up. This is when a study is said to be irreproducible. The second step is when it becomes apparent—if it does—*why* the original work was irreproducible. If it is because of fabricated data, a career is probably over. If it is because of bad technique, a reputation is on the line. Yet it is important to understand here that failure to replicate a study is not *necessarily* a bad thing for the profession at large. This is how science learns.[75] This is probably in part what is meant by saying that science is self-correcting. A scientist makes a mistake, another scientist catches it, and a lesson is learned. If the lesson is about the object of inquiry—rather than the ethical standards of the original researcher—further work is possible. Perhaps even enlightenment. When we think that we know something, but really don't, it is a barrier to further discovery. But when we find that something we thought was true really wasn't, a breakthrough may be ahead. We have the opportunity to learn more about how things work. Thus failure can be valuable to science.[76]

The most important thing about science is that we *try* to find failure. The real danger to science comes not from mistakes but deception. Mistakes can be corrected and learned from; deception is often used to cover up mistakes. Fabricated data may be built upon by scores of other researchers before it is caught. When individual researchers put their own careers ahead of the production of knowledge, they are not only cheating on the ideals of the scientific attitude, they are cheating their professional colleagues as well. Lying, manipulation, and tampering are the very definition of having the wrong attitude about empirical evidence.

But again, we must make every effort to draw a distinction between irreproducible studies and fraud. It is helpful here to remember our logic. Fraudulent results are almost always irreproducible, but this does not mean that all or even most irreproducible studies are the result of fraud. The crime against the scientific attitude is of course much greater in cases of fraud. Yet we can also learn something about the scientific attitude by focusing on some of the less egregious reasons for irreproducible studies. Even if the reason for a study not being replicated has *nothing whatsoever to do with fraud,* scrutiny by the larger scientific community will probably find it. Merely by having a standard of data sharing and reproducibility, we are embracing the scientific attitude.[77] Whether an irreproducible study results from intentional or unintentional mistakes, we can count on the same mechanism to find them.

One might think, therefore, that most scientific studies are replicated. That at least half of all scientific work must involve efforts to reproduce others' findings so that we can be sure they are right before the field moves forward. This would be false. Most journals do not welcome publication of studies that merely replicate someone else's results. It is perhaps more interesting if a study *fails* to replicate someone else's result, but here too there is some risk. The greater glory in scientific investigation comes from producing one's own original findings, not merely checking someone else's, whether they can be replicated or not. Perhaps this is why only a small percentage of scientific work is even *attempted* to be replicated.[78] One might argue that this is appropriate because the standards of peer review are so high that one would not expect most studies to *need* to be repeated. With all that scrutiny, it must be a rare event to find a study that is irreproducible. But this assumption too has been challenged in recent years. In fact, the media in the last few years have been full of stories about the "reproducibility crisis" in science, especially in psychology, where—when someone bothered to look—researchers found that 64 percent of studies were irreproducible![79]

Like the p-hacking "crisis," one might see this as a serious challenge to the claim that science is special, and to the claim that scientists are distinctive in their commitment to a set of values that honor basing one's beliefs on empirical evidence. But again, much can be learned by examining how the scientific community has responded to this replication "crisis."

Let's start with a closer look at the crisis itself. In August 2015, Brian Nosek and colleagues published a paper entitled "Estimating the Reproducibility of Psychological Science," in the prestigious journal *Science*. It dropped like a bombshell. A few years earlier Nosek had cofounded the Center for Open Science (COS), which had as its mission to increase data sharing and replication of scientific experiments. Their first project was entitled the Reproducibility Project. Nosek recruited 270 other researchers to help him attempt to reproduce one hundred psychological studies. And what they found was shocking. Out of one hundred studies, only 36 percent percent of the findings could be reproduced.[80]

It is important to point out here that Nosek et al. were not claiming that the irreproducible studies were fraudulent. In every case they found an effect in the same direction that the original researchers had indicated. It's just that in most of them the effect size was cut in half. The evidence was not as strong as the original researchers had said it was, so the conclusion of the original papers was unsupported in over half of the cases. Why did this happen? One's imagination immediately turns to all of the "degrees of freedom" that were discussed earlier in this chapter. From too small a sample size to analytical error, the replicators found a number of different shortcomings.

This led to headline announcements throughout the world heralding a replication crisis in science. And make no mistake, the fallout was not just for psychology. John Ioannidis (whom we remember for his earlier paper "Most Research Results Are Wrong") said that the problem could be even worse in other fields, such as cell biology, economics, neuroscience, or clinical medicine.[81] "Many of the biases found in psychology are pervasive," he said.[82] Others agreed.[83] There were, of course, some skeptics. Norbert Schwarz, a psychologist from the University of Southern California, said "there's no doubt replication is important, but it's often just an attack, a vigilante exercise." Although he was not involved in any of the original studies, he went on to complain that replication studies themselves are virtually never evaluated for errors in their *own* design or analysis.[84]

Yet this is exactly what happened next. In March 2016, three researchers from Harvard University (Daniel Gilbert, Gary King, and Stephen Pettigrew) and one of Nosek's colleagues from the University of Virginia (Timothy Wilson), released their own analysis of the replication studies put forward by the Center for Open Science and found that "the research methods used to reproduce those studies were poorly designed, inappropriately applied and introduced statistical error into the data."[85] When a re-analysis of the original studies was done using better technique, the reproducibility rate was close to 100 percent. As Wilson put it, "[the Nosek study] got so much press, and the wrong conclusions were drawn from it.... It's a mistake to make generalizations from something that was done poorly, and this we think was done poorly."[86]

This is a breathtaking moment for those who care about science. And, I would argue, this is one of the best demonstrations of the critical attitude behind science that one could imagine. A study was released that found flawed methodology in some studies, and—less than seven months later—it was revealed that there were flaws in the study that found the flaws. Instead of a "keystone cops" conclusion, it is better to say that in science we have a situation where the gatekeepers also have gatekeepers—where there is a system of checking and rechecking in the interest of correcting errors and getting closer to the truth.

What were the errors in Nosek's studies? They were numerous.

(1) Nosek et al took a nonrandom approach in selecting which studies to try to replicate. As Gilbert explains, "what they did is create an idiosyncratic, arbitrary list of sampling rules that excluded the majority of psychology subfields from the sample, that excluded entire classes of studies whose methods are probably among the best in science from the sample.... Then they proceeded to violate all of their own rules.... So the first thing we realized was that no matter what they found—good news or bad news—they never had any chance of estimating the reproducibility of psychological science, which is what the very title of their paper claims they did."[87]

(2) Some of their studies were not even close to exact replications. In one of the most egregious examples, the COS team were trying to recreate a study on attitudes toward affirmative action that had been done at Stanford. The original study had asked white students at Stanford to watch a video of four other Stanford students—three white and one black—talking about race, during which one of the white students made an offensive comment. It was found that the observers looked significantly longer at the black student when they believed that he could hear the comment than when they believed that he could not. In their attempt at replication, however, the COS team showed the video to students at… the University of Amsterdam. Perhaps it is not shocking that Dutch students, watching American students speaking in English, would not have the same reaction. But here is where the real trouble came, for when the COS team realized the problem and reran the experiment at a different American university and got the same result as the original Stanford study, they chose to leave this successful replication out of their report, but included the one from Amsterdam that had failed.[88]

(3) There was no agreed-upon protocol for quantitative analysis before the replication. Instead, Nosek and his team used five different measures (including strength of the effect) to look at all of the results combined. It would have been better, Gilbert et al. recommended, to focus on one measure.[89]

(4) The COS team failed to account for how many of the replications would have been expected to fail *purely by chance*. It is indefensible statistical method to expect a 100 percent success rate in replications. King explains, "If you are going to replicate 100 studies, some will fail by chance alone. That's basic sampling theory. So you have to use statistics to estimate how many studies are expected to fail by chance alone because otherwise the number that actually do fail is meaningless."[90]

The result of all these errors is that—if you take them into account and undo them—the reproducibility rate for the original one hundred studies was "about what we should expect if every single one of the original findings had been true."[91] It should be noted here that even so, Gilbert, King, Pettigrew, and Wilson are not suggesting fraud or any other type of malfeasance on the part of Nosek and his colleagues. "Let's be clear," Gilbert said, "No one involved in this study was trying to deceive anyone. They just made mistakes, as scientists sometimes do."[92]

Nosek nonetheless complained that the critique of his replication study was highly biased: "They are making assumptions based on selectively interpreting data and ignoring data that's antagonistic to their point of view."[93] In short, he accused them almost precisely of what they had accused him of doing—cherry picking—which is what he had accused some of the original researchers of doing too.

Who is right here? Perhaps they all are. Uri Simonsohn (one of the coauthors of the previously discussed p-hacking study by Simmons et al.)—who had absolutely nothing to do with this back-and-forth on the replication crisis and its aftermath—said that both the original replication paper and the critique used statistical techniques that are "predictably imperfect." One way to think about this, Simonsohn said, is that Nosek's paper said the glass was 40 percent full, whereas Gilbert et al. said it could be 100 percent full.[94] Simonsohn explains, "State-of-the-art techniques designed to evaluate replications say it is 40 percent full, 30 percent empty, and the remaining 30 percent could be full or empty, we can't tell till we get more data."[95]

What are the implications for the scientific attitude? They are manifest in this last statement. When we don't have an answer to an empirical question, we must investigate further. Researchers used the scientific attitude to critique the study that critiqued other studies. But even here, we have to check the work. It is up to the scientific community as a whole to decide—just as they did in the case of cold fusion—which answer is ultimately right.

Even if all one hundred of the original studies that were examined by the Reproducibility Project turn out to be right, there needs to be more scrutiny on the issues of data sharing and replication in science. This dust up—which is surely far from over—indicates the need to be much more transparent in our research and reporting methods, and in trying to replicate one another's work. (Indeed, look how valuable the reproducibililty standard was in the cold fusion dispute.) And already perhaps some good has come of it. In 2015, the journal *Psychological Science* announced that it will now ask researchers to preregister their study methods and modes of analysis prior to data collection, so that it can later be reconciled with what they actually found. Other journals have followed suit.[96]

The increased scrutiny that the "reproducibility crisis" has created is a good thing for science. Although it is surely embarrassing for some of its participants, the expectation that nothing should be swept under the rug marks yet another victory for the scientific attitude. Indeed, what other field can we imagine rooting out its own errors with such diligence? Even in the face of public criticism, science remains completely committed to the highest standards of evidence. While there may be individual transgressions, the "replication crisis" in science has made clear its commitment to the scientific attitude.

Conclusion

In this chapter, I've come down pretty hard on some of the mistakes of science. My point, however, was not to show that science is flawed (or that it is perfect), even if I did along the way show that scientists are human. We love our own theories. We want them to be right. We surely make errors of judgment all the time in propping up what we hope to be true, even if we are statisticians. My point has not been to show that scientists do not suffer from some of the same cognitive biases, like confirmation bias and motivated reasoning, as the rest of us. Arguably, these biases may even be the reason behind p-hacking, refusal to share one's data, taking liberties with degrees of freedom, running out to a press conference rather than subjecting oneself to peer review, and offering work that is irreproducible (even when our subject of inquiry is irreproducible work itself). Where the reader might have been tempted in various cases to conclude that such egregious errors could only be the result of intentional oversight—and must therefore reveal fraud—I ask you to step back and consider the possibility that they may be due merely to unconscious cognitive bias.[97]

Fortunately, science *as an institution* is more objective than its practitioners. The rigorous methods of scientific oversight are a check against individual bias. In several examples in this chapter we saw problems that were created by individual ambition and mental foibles that could be rectified through the rigorous application of community scrutiny. It is of course hoped that the scientific attitude would be embraced by all of its individual practitioners. But as we've seen, the scientific attitude is more than just an individual mindset; it is a shared ethos that is embraced by the community of scholars who are tasked with judging one another's theories against publicly available standards. Indeed, this may be the real distinction between science and pseudoscience. It is not that pseudoscientists suffer from more cognitive bias than scientists. It is not even that scientists are more rational (though I hope this is true). It is instead that *science has made a community-wide effort to create a set of evidential standards that can be used as a check against our worst instincts and make corrections as we go, so that scientific theories are warranted for belief by those beyond the individual scientists who discovered them.* Science is the best way of finding and correcting human error on empirical matters not because of the unusual honesty of scientists, or even their individual commitment to the scientific attitude, but because the mechanisms for doing so (rigorous quantitative method, scrutiny by one's peers, reliance on the refutatory power of evidence) are backed up by the scientific attitude at the community level.

It is the scientific *attitude*—not the scientific *method*—that matters. As such, we need no longer pretend that science is completely "objective" and that "values" are irrelevant. In fact, values are crucial to what is special about science. Kevin deLaplante says, "Science is still very much a 'value-laden' enterprise. Good science is just as value-laden as bad science. What distinguishes good from bad science isn't the absence of value judgments, it's the kind of value judgments that are in play."[98]

There are still a few problems to consider. First, what should we do about those who think that they have the scientific attitude when they don't? This concern will be dealt with in chapter 8, where we will discuss the problems of pseudoscience and science denialism. I trust, however, that one can already begin to imagine the outlines of an answer: it is not just an individual's reliance on the scientific attitude—and surely not his or her possibly mistaken belief about fulfilling it—that keeps science honest. The community will judge what is and is not science. Yet now a second problem arises, for is it really so easy to say that the "community decides"? What happens when the scientific community gets it wrong? In chapter 8, I will also deal with this problem, via a consideration of those instances where the individual scientist—like Galileo—was right and the community was wrong.[99] As an example I will discuss the delightful, but less well-known, example of Harlen Bretz's theory that a megaflood was responsible for creating the eastern Washington State scablands. I will also offer some reasons for disputing the belief that today's climate change deniers may just be tomorrow's Galileo. And, as promised, I will examine the issue of scientific fraud in chapter 7.

Before all that, however, in the next chapter, we will take a break from consideration of scientific drawbacks and failures to take a closer look at one example of an unabashed success of the scientific attitude and how it happened: modern medicine.

6 How the Scientific Attitude Transformed Modern Medicine

It is easy to appreciate the difference that the scientific attitude can make in transforming a previously undisciplined field into one of scientific rigor, for we have the example of modern medicine. Prior to the twentieth century, the practice of medicine was based largely on hunches, folk wisdom, and trial and error. Large-scale experiments were unknown and data were difficult to gather. Indeed, even the idea that one needed to test one's hypotheses against empirical evidence was rare. All of this changed within a relatively short period of time after the germ theory of disease in the 1860s and its translation into clinical practice in the early twentieth century.[1]

We already saw in chapter 1 how Ignaz Semmelweis's discovery of the cause of childbed fever in 1846 provides a prime example of what it means to have the scientific attitude. We also saw that he was far ahead of his time and that his ideas were met with unreasoned opposition. The scientific attitude that Semmelweis embraced, however, eventually found fruition throughout medicine. At about the same time as Semmelweis's work, medicine saw the first public demonstration of anesthesia. For the first time, surgeons could take their time doing operations, as they no longer had to wrestle down fully awake patients who were screaming in pain. This did not in and of itself allow for lower mortality, as one complicating factor from lengthened surgeries was that patients also had more time for their wounds to be open to the air and get infected.[2] Only after Pasteur discovered bacteria and Koch detailed the process of sterilization did the germ theory of disease begin to take root. When Lister introduced antiseptic techniques (which kill the germs) and aseptic surgery (which prevents the germs from entering in the first place) in 1867, it was finally possible to keep the cure from sometimes being worse than the disease.[3]

From today's point of view, it is easy to take these advances for granted and underappreciate how they led to the growth of better quantitative techniques, laboratory analysis, controlled experimentation, and the idea that diagnosis and treatment should be based on evidence rather than intuition. But one should not forget that Western medicine has *always* fancied itself to be scientific; it is just that the meaning of the term has changed.[4] Astrological medicine and bloodletting were once considered cutting edge, based on the highest principles of rationality and experience. One would be hard pressed to find any eighteenth-century doctor—or I imagine even one in the early Greek era—who did not consider his knowledge "scientific."[5] How can one claim, then, that these early physicians and practitioners were so woefully benighted? As we have seen, such things are judged by the members who make up one's profession and so are relative to the standards of the age—but according to the standards of the early nineteenth-century, bloodletting seemed just fine.

My goal in this chapter is not to disparage the beliefs of any particular period, even as they cycled through untested beliefs to outrageous hypotheses, sometimes killing their patients with what we would today recognize as criminal incompetence. Instead, I would like to shine a light on how medicine found its way out of these dark ages and moved toward a time when its practice could be based on careful observation, calculation, experiment, and the flexibility of mind to accept an idea when (and only when) it had been empirically demonstrated. Remember that the scientific attitude requires not just that we care about evidence (for what counts as evidence can change from age to age) but that we are *willing to change our theory based on new evidence*. It is this last part that is crucial when some area of inquiry is trying to make the jump from pseudoscience to science, from mere opinion to warranted belief. For it was only when this mindset was finally embraced—when physicians stopped thinking that they already had all the answers based on the authority of tradition and began to realize that they could learn from experiment and from the experience of others—that medicine was able to come forward as a science.

The Barbarous Past

In his excellent book *The History of Medicine: A Very Short Introduction*, William Bynum reminds us that—despite bloodletting, toxic potions, skull drilling, and a host of "cures" that were all too often worse than the disease—Western medicine has always considered itself modern.[6]

One of the greatest systems of medicine in antiquity came from Hippocrates, whose chief insight (besides the Hippocratic Oath for which he is famous) involved his theory of how the "four humors" could be applied to medicine. As Roy Porter puts it in his masterful work *The Greatest Benefit to Mankind*:

> From Hippocrates in the fifth century BC through to Galen in the second century AD, "humoral medicine" stressed the analogies between the four elements of external nature (fire, water, air and earth) and the four humours or bodily fluids (blood, phlegm, choler or yellow bile and black bile), whose balance determined health.[7]

The theory was impressive and the explanatory potential intriguing. Since disease was thought to be due to an imbalance in the humors—colds were due to phlegm, vomiting was due to bile, and so on—health could be maintained by keeping them in balance, for instance through the practice of bloodletting.[8]

Though bloodletting was invented by Hippocrates, Galen brought it to a high art, and for more than a thousand years thereafter (right through the nineteenth century) it was seen as a therapeutic treatment. Having written four separate books on the pulse, Galen thought that bloodletting allowed the healer to capitalize on the example of nature, where the removal of excess fluid—such as during menstruation—prevented disease.[9] As Porter puts it, "whatever the disorder—even blood loss—Galen judged bleeding proper."[10] As such, he often bled his patients to unconsciousness (which sometimes resulted in their death).

This is not the only ancient medical practice that we would judge barbarous from today's point of view; there was also skull drilling, leeches, swallowing mercury, the application of animal dung, and more. What is shocking, however, is the extent to which such ignorance went unchallenged down through the centuries, with the result that until quite recently in the history of medicine, patients often had just as much to fear from their doctor as they did from whatever disease they were trying to cure.

It is not just that the theories in circulation at the time were wrong—for, as we have seen, many scientific theories will turn out to be mistaken—but that most of these ideas were not even based on any sort of evidence or experiment in the first place. Medicine did not yet have the scientific attitude.

The Dawn of Scientific Medicine

The transition out of this nonempirical phase of medicine was remarkably slow. The Scholastic tradition lingered in medicine long after it had been abandoned by astronomy and physics, with the result that even two hundred years after the scientific revolution had begun in the seventeenth century, medical questions were customarily settled by theory and argument—to the extent that they were settled at all—rather than controlled experiment.[11] Both the empirical and clinical practices of medicine remained quite backward until the middle of the nineteenth century. Even to the extent that a fledgling science of medicine began to emerge during the Renaissance, it had more of an effect on knowledge than on health.[12] Despite the great breakthroughs in anatomy and physiology in early-modern times (e.g., Harvey's seventeenth-century work on the circulation of blood), "[medicine's] achievements proved more impressive on paper than in bedside practice."[13] In fact, even the one unambiguous improvement in medical care in the eighteenth century—the vaccine against smallpox—is seen by some not as the result of science so much as embracing popular folk wisdom.[14] This "nonscientific" outlook persisted in medicine straight through the eighteenth century (which was known as the "age of quackery"); it was not until the middle of the nineteenth century that modern medicine truly began to arise.[15]

In his delightful memoir, *The Youngest Science*, Lewis Thomas contrasts the kind of scientific medicine practiced today with

> the kind of medicine taught and practiced in the early part of the nineteenth century, when anything that happened to pop into the doctor's mind was tried out for the treatment of illness. The medical literature of those years makes horrifying reading today: paper after learned paper recounts the benefits of bleeding, cupping, violent purging, the raising of blisters by vesicant ointments, the immersion of the body in either ice water or intolerably hot water, endless lists of botanical extracts cooked up and mixed together under the influence of nothing more than pure whim. ... Most of the remedies in common use were more likely to do harm than good.[16]

Bloodletting in particular seemed popular, owing in part to its enthusiastic support by the prominent physician Benjamin Rush. A holdover from premedieval times, bloodletting was thought to have enormous health benefits and constituted one of the "extreme interventions" that were thought necessary for healing. In his important book *Seeking the Cure,* Dr. Ira Rutkow writes that

> citizens suffered needlessly as a result of Rush's egotism and lack of scientific methodologies. In an age when no one understood what it meant to measure blood pressure and body temperature, and physicians were first determining the importance of heart and breathing rates, America's doctors had no parameters to prevent them from harming patients.[17]

Yet harm them they did.

> Doctors bled some patients sixteen ounces a day up to fourteen days in succession.... Present-day blood donors, by comparison, are allowed to give one pint (sixteen ounces) per session, with a minimum of two months between each donation. Nineteenth-century physicians bragged about the totality of their bleeding triumphs as if it were a career-defining statistic.... The regard for bloodletting was so deep-seated that even the frequent complications and outright failures to cure did not negate the sway of Rush's work.[18]

Fittingly Rush died, in 1813, as a result of treatment by bloodletting for his typhus fever.[19] Yet this was not the only horrific practice of the time.

> Medicine was performed literally in the dark. Electricity was newfangled and unpopular. Almost every act a doctor performed—invasive examinations, elaborate surgeries, complicated births—had to be done by sun or lamplight. Basics of modern medicine, such as the infectiousness of diseases, were still under heavy dispute. Causes of even common diseases were confusing to doctors. Benjamin Rush thought yellow fever came from bad coffee. Tetanus was widely thought to be a reflex irritation. Appendicitis was called peritonitis, and its victims were simply left to die. The role that doctors—and their unwashed hands and tools—played in the spread of disease was not understood. "The grim spectre of sepsis" was ever present. It was absolutely expected that wounds would eventually fester with pus, so much so that classifications of pus were developed.... Medicine was not standardized, so accidental poisoning was common. Even "professionally" made drugs were often bulky and nauseating. Bleeding the ill was still a widespread practice, and frighteningly large doses of purgatives were given by even the most conservative men. To treat a fever with a cold bath would have been "regarded as murder." There was no anesthesia—neither general nor local. Alcohol was commonly used when it came to enduring painful treatments... and pure opium [was] sometimes available too. If you came to a doctor for a compound fracture, you had only a fifty percent chance of survival. Surgery on brains and lungs was attempted only in accident cases. Bleeding during operations was often outrageously profuse, but, as comfortingly described by one doctor, "not unusually fatal."[20]

It is important to contrast what was occurring in the United States (which was fairly late to the scientific party in medicine) with progress that was already underway in Europe. In the early nineteenth century, Paris in particular was a center for many advances in medical understanding and practice, owing perhaps to the revolutionary outlook that had seen hospitals move out of the hands of the church to become nationalized.[21] In Paris, a more empirical outlook prevailed. Autopsies were used to corroborate bedside diagnosis. Doctors appreciated the benefits of palpation, percussion, and auscultation (listening), and René Laennec invented the stethoscope. A more naturalistic outlook in general brought medical students to learn at their patients' bedsides in hospitals. Following these developments, the rise of scientific labs in Germany, and appreciation for the microscope by Rudolf Virchow and others, led to more advances in basic science.[22] All of this activity attracted medical students from around the world—particularly the United States—to come to France and Germany for a more scientifically based medical education. Even so, the benefits of this for direct medical practice were slow to arise on either side of the Atlantic. As we have seen, Semmelweis was one of the earliest physicians to try to bring a more scientific attitude to patient care in Europe at roughly the same time the miracle of anesthesia was being demonstrated in Boston. Yet even these advances met with resistance.[23]

The real breakthrough arrived in the 1860s with the discovery of the germ theory of disease. Pasteur's early work was on fermentation and the vexed question of "spontaneous generation." In his experiments, Pasteur sought to show that life could not arise out of mere matter. But why, then, would a flask of broth that had been left open to the air "go bad" and produce organisms?[24]

> Pasteur devised an elegant sequence of experiments. He passed air through a plug of gun-cotton inserted into a glass tube open to the atmosphere outside his laboratory. The gun-cotton was then dissolved and microscopic organisms identical to those present in fermenting liquids were found in the sediment. Evidently the air contained the relevant organisms.[25]

In later experiments, Pasteur demonstrated that these organisms could be killed by heat. It took a few more years for Pasteur to complete more experiments involving differently shaped flasks placed in different locations, but by February 1878 he was ready to make the definitive case for the germ theory of infection before the French Academy of Medicine, followed by a paper that made the case that microorganisms were responsible for disease.[26] The science of bacteriology had been born.

Lister's work on antisepsis—which aimed at preventing these microorganisms from infecting the wounds caused during surgery—grew directly out of Pasteur's success.[27] Lister had been an early adopter of Pasteur's ideas, which were still resisted and misunderstood throughout the 1870s.[28] Indeed, when US president James Garfield was felled by an assassin's bullet in 1871, he died many months later not, many felt, as a result of the bullet that was still lodged in his body, but the probing of the wound site with dirty fingers and instruments by some of the most prominent physicians of the time. At trial, Garfield's assassin even tried to defend himself by saying that the president had died not of being shot but of medical malpractice.[29]

Robert Koch's laboratory work in Germany during the 1880s took things the next step. Skeptics had always asked about germs, "Where are the little beasts?"—not accepting the reality of something they could not see.[30] Koch finally answered this with his microscopic work, where he was able not only to establish the physical basis for the germ theory of disease but even to identify the microorganisms responsible for specific diseases.[31] This led to the "golden years" of bacteriology (1879–1900), when "the microorganisms responsible for major diseases were being discovered at the phenomenal rate of one a year."[32]

All of this success, though, may now compel us to ask a skeptical question. If this was such a golden age of discovery in bacteriology—that provided such stunning demonstration of the power of careful empirical work and experiment in medicine—why did it not have a more immediate effect on patient care? Surely, there were some timely benefits (for instance, Pasteur's work on rabies and anthrax), but one nonetheless feels a lack of proportion between the good science being done and its effect on treatment. If we here see the beginning of respect for empirical evidence in medical research, why was there such a lag for the realization of its fruit in clinical science?

> As of the late nineteenth century the few medicines that were effective included mercury for syphilis and ringworm, digitalis to strengthen the heart, amyl nitrate to dilate the arteries in angina, quinine for malaria, colchicum for gout—and little else.... Blood-letting, sweating, purging, vomiting and other ways of expelling bad humours had a hold upon the popular imagination and reflected medical confidence in such matters. Blood-letting gradually lost favour, but it was hardly superseded by anything better.[33]

Perhaps part of the reason was that during this period experiment and practice tended to be done by different people. Social bifurcation within the medical community tended to enforce an unwritten rule whereby researchers didn't practice and practitioners didn't do research. As Bynum says in *The Western Medical Tradition*:

> "Science" had different meanings for different factions within the wider medical community. … "Clinical science" and "experimental medicine" sometimes had little to say to each other. … These two pursuits were practised increasingly by separate professional groups. … Those who produced knowledge were not necessarily those who used it.[34]

Thus the scientific revolution that had occurred in other fields (physics, chemistry, astronomy) two hundred years earlier—along with consequent debates about methodology and the proper relationship between theory and practice—did not have much of an effect on medicine.[35] Even after the "bacteriological revolution," when the knowledge base of medicine began to improve, the lag in clinical practice was profound.

> Debates about the methodology of natural philosophy [science] touched medicine only obliquely. Medicine continued to rely on its own canonical texts, and had its own procedures and sites for pursuing knowledge: the bedside and the anatomy theatre. Most doctors swore by the tacit knowledge at their fingers' ends.[36]

For all its progress, medicine was not yet a science. Even if all of this hard-won new understanding was now available for those who cared to pursue it, a systemic problem remained: how to bridge the gap between knowledge and healing in those who were practicing medicine, and how to ensure that this new knowledge could be put into the hands of those students who were the profession's future?[37]

Medicine at this point faced an organizational problem not just for the creation but the transmission of new knowledge. Instead of a well-oiled machine for getting scientific breakthroughs into the hands of those who could make best use of them, at this time medicine was a "contested field occupied by rivals."[38] Although some of the leading lights in medicine had already embraced the scientific attitude, the field as a whole still awaited its revolution.

The Long Transition to Clinical Practice

For whatever reason—ideological resistance, ignorance, poor education, lack of professional standards, distance between experimenters and practitioners—it was a long transition to the use of the science of medicine in clinical practice. Indeed, some of the horror stories of untested medical treatments and remedies from the nineteenth century held over right through the early part of the twentieth century. In the United States, there continued to be poor physician education and lack of professional standards by which practitioners could be held accountable for their sometimes-shoddy practices.

At the dawn of the twentieth century, physicians still did not know enough to cure most diseases, even if they were now better at identifying them. With the acceptance of the germ theory of disease, physicians and surgeons were now perhaps not as likely to kill their patients through misguided interventions as they had been in earlier centuries, but there was still not much that the scientific breakthroughs of the 1860s and 1870s could do for direct patient care. Lewis Thomas writes that

> explanation was the real business of medicine. What the ill patient and his family wanted most was to know the name of the illness, and then, if possible, what had caused it, and finally, most important of all, how it was likely to turn out.... For all its facade as a learned profession, [medicine] was in real life a profoundly ignorant profession.... I can recall only three or four patients for whom the diagnosis resulted in the possibility of doing something to change the course of the illness.... For most of the infectious diseases on the wards of the Boston City Hospital in 1937, there was nothing to be done beyond bed rest and good nursing care.[39]

James Gleick paints a similarly grim portrait of medical practice in this era:

> Twentieth-century medicine was struggling for the scientific footing that physics began to achieve in the seventeenth century. Its practitioners wielded the authority granted to healers throughout human history; they spoke a specialized language and wore the mantle of professional schools and societies; but their knowledge was a pastiche of folk wisdom and quasi-scientific fads. Few medical researchers understood the rudiments of controlled statistical experimentation. Authorities argued for or against particular therapies roughly the way theologians argued for or against their theories, by employing a combination of personal experience, abstract reason, and aesthetic judgment.[40]

But change was coming, and a good deal of it was social. With advances in anesthesia and antisepsis, joined by the bacteriological discoveries of Pasteur and Koch, by the early twentieth century clinical medicine was ripe to emerge from its unenlightened past. As word of these discoveries spread, the diversity of treatments and practices in clinical medicine suddenly became an embarrassment. Of course, as with any paradigm shift, Kuhn has taught us that one of the most powerful forces at work is that the holdouts die and take their old ideas with them, while new ideas are embraced by younger practitioners. Surely some of this occurred (Charles Meigs, one of the staunchest resisters of Semmelweis's theory on childbed fever *and* anesthesia, died in 1869), but social forces were likely even more influential.

Modern medicine is not just a science but also a social institution. And it is important to realize that even before it was truly scientific, social forces shaped how medicine was viewed and practiced, and had a hand in turning it into the science that it is today. Entire books have been written on the social history of medicine; here I will have the chance to tell only part of the story.

In *The Social Transformation of American Medicine*, Paul Starr argues that the democratic principles of early America conflicted with the idea that medical knowledge was somehow privileged and that the practice of medicine should be left to an elite.[41] In its earliest days, "all manner of people took up medicine in the colonies and appropriated the title of doctor."[42] As medical schools began to open, and those who had received more formal medical training sought to distance themselves from the "quacks" by forming medical societies and pursuing licensing, one might think that this would have been welcomed by a populace eager for better medical care. But this did not occur. Folk medicine and lay healing continued to be popular, as many viewed the professionalization movement in American medicine as an attempt to grab power and authority. As Starr writes,

> Popular resistance to professional medicine has sometimes been portrayed as hostility to science and modernity. But given what we now know about the objective ineffectiveness of early nineteenth-century therapeutics, popular skepticism was hardly unreasonable. Moreover, by the nineteenth century in America, popular belief reflected an extreme form of rationalism that demanded science be democratic.[43]

In fact, to the extent that medical licensing standards had already begun to be established in early America, by the time of Andrew Jackson's presidency in the 1830s, there was an organized effort to have them disbanded as "licensed monopolies."[44] Incredibly, this led to the abandonment of medical licensing standards in the United States for the next fifty years.[45]

To anyone who cares about the science of medicine—not to mention the effect that all of this might have had on patient care—this is a sorry result. Of course, one understands that, given the shoddy medical practices of the time, there was deep suspicion over whether "professional" physicians had any better training or knew anything more about medical care than lay practitioners did.[46] Still, the demand that scientific knowledge be "democratic" can hardly be thought of as a promising development for better patient care, when the scientifically based medical discoveries of Europe were still waiting to take root in clinical practice. The result was that by the end of the nineteenth century—just as a revolution in basic medical knowledge was taking place in Europe—the state of US medical education and practice was shameful.

Medical "diploma mills," which were run for profit by local physicians but offered no actual instruction in the basic sciences or hands-on training in patient care, were rampant.[47] The restoration of medical licensing in the 1870s and 1880s led to a little more professional accountability, as the idea that one's diploma was the only license needed to practice medicine came under scrutiny. Eventually, the requirements stiffened.

> One major landmark was an 1877 law passed by Illinois, which empowered a state board of medical examiners to reject diplomas from disreputable schools. Under the law, all doctors had to register. Those with degrees from approved schools were licensed, while others had to be examined. Of 3,600 nongraduates practicing in Illinois in 1877, 1,400 were reported to have left the state within a year. Within a decade, three thousand practitioners were said to have been put out of business.[48]

In Europe, the standards of medical education were higher. German medical schools in particular offered training in schools that were affiliated with actual universities (as few did in America). Eventually, this led to the emulation of a more rigorous model of medical education in the United States with the founding of Johns Hopkins Hospital in 1889, and its affiliated medical school four years later, which included instruction at all levels, including internships and residencies.[49] The medical schools at Harvard, Hopkins, Penn, Michigan, Chicago, and a few others in the United States that were affiliated with universities were widely respected.[50] But this accounted for only a fraction of medical education at the end of nineteenth-century America.

In 1908, a medical layperson named Abraham Flexner took off on a quest—under the aegis of the Carnegie Foundation and the Council on Medical Education of the AMA—to visit all 148 American medical schools then in existence. What he found was appalling.

> Touted laboratories were nowhere to be found, or consisted of a few vagrant test tubes squirreled away in a cigar box; corpses reeked because of the failure to use disinfectant in the dissecting rooms. Libraries had no books; alleged faculty members were busily occupied in private practice. Purported requirements for admission were waived for anyone who would pay the fees.[51]

In one particularly colorful example, Flexner visited a medical school in Des Moines, Iowa, where the dean rushed him through his visit. Flexner had seen the words "anatomy" and "physiology" stenciled on doors, but they were all locked and the dean told him that he did not have the keys. Flexner concluded his visit, then doubled back and paid a janitor to open the doors, where he found that every room was identical, having only desks, chairs, and a small blackboard.[52]

When Flexner's famous report came out in 1910 it was an indictment of the vast majority of medical education in the United States. Johns Hopkins was held up as the gold standard, and even other reputable schools were encouraged to follow its model, but virtually all of the commercial schools were found to be inadequate. Flexner argued, among other reforms, that medical schools needed to be rooted in an education in the natural sciences, and that reputable medical schools should be affiliated with a university. They also needed to have adequate scientific facilities. He further recommended that students should have at least two prior years of college before starting medical training, that medical faculty should be full time, and that the number of medical schools should be reduced.[53]

The effect was immediate and profound.

> By 1915 the number of [medical] schools had fallen from 131 to 95, and the number of graduates from 5,440 to 3,536.... In five years, the schools requiring at least one year of college work grew from thirty-five to eighty-three.... Licensing boards demanding college work increased from eight to eighteen. In 1912 a number of boards formed a voluntary association, the Federation of State Medical Boards, which accepted the AMA's rating of medical schools as authoritative. The AMA Council effectively became a national accrediting agency for medical schools, as an increasing number of states adopted its judgments of unacceptable institutions.... [By 1922] the number of medical schools had fallen to 81, and its graduates to 2,529. Even though no legislative body ever set up either the Federation of State Medical Boards or the AMA Council on Medical Education, their decisions came to have the force of law. This was an extraordinary achievement for the organized profession.[54]

As they took over licensing requirements, state medical boards began to have much more power, not only in their oversight of medical education but in sanctioning practitioners who were already in the field. With the creation of the FSMB, there was now a mechanism in place not only to bring more evidence-based medicine into the training of new physicians, but also to hold existing physicians accountable for their sometimes-shoddy practices. In 1921, the American College of Surgeons released its minimum standards of care and there was a push for hospitals to become accredited.[55] This did not mean that every state suddenly had the legal teeth to root out questionable practitioners (as occurred in Illinois as early as 1877). But it did mean at least that the most egregious practices (and practitioners) were now under scrutiny. Even if there was still not a lot that most physicians could do to cure their patients, they could be ostracized for engaging in practices that harmed them. In some states, the names of bad physicians were even reported in state medical bulletins.

Within just a few years after the Flexner Report, the scientific revolution in medicine that had started in Europe in the 1860s had finally come to the United States. Although the majority of this change was social and professional rather than methodological or empirical, the sum effect was that by excluding those physicians without adequate training, and cracking down on practices that were no longer acceptable, a set of social changes that may have started off in self-interest and the protection of professional standing resulted in furthering the sort of *group scrutiny* of *individual practices* that is the hallmark of the scientific attitude.

Of course, it may still have been true, as Lewis Thomas's previously cited portrait makes clear, that even at the best hospitals in Boston in the 1920s, there was little that most doctors could do for their patients beyond homeopathic drugs (which were placebos) or surgery, other than wait for the disease to run its natural course. Even if they were no longer bleeding, purging, blistering, cupping (and killing) their patients—or probing them with dirty fingers and instruments—there were as yet few direct medical interventions that could be offered to heal them. Yet this nonetheless represented substantial progress over an earlier era. Medicine finally embraced the beginning of professional oversight of individual practices that was necessary for it to come forward as a science. Medical knowledge was beginning to be based on empirical evidence, and the introduction of standards of care promised at least that clinical science would make a good faith effort to live up to this (or at least not undermine it). Medicine was no longer based on mere hunches and anecdotes. Bad practices and ineffective treatments could be scrutinized and discarded. By raising its professional standards, medicine had at last lived up to its scientific promise.

This is the beginning of the scientific attitude in medicine. One could make the case that the reliance on empirical evidence and its influence on theory went all the way back to Semmelweis or even Galen.[56] An even better case could probably be made for Pasteur. As in any field, there were giants throughout the early history of medicine, and these tended to be those who embraced the idea of learning from empirical evidence. But the point about the importance of a community ethos still stands, for if an entire field is to become a science, the scientific attitude has to be embraced by more than just a few isolated individuals, no matter how great. One can point to individual examples of the scientific attitude in early medicine, but it was not until those values were widespread in the profession—at least in part because of social changes in the medical profession itself—that one can say that medicine truly became a science.

The Fruits of Science

After the professional reforms of the early twentieth century, medicine came into its own. With the discovery of penicillin in 1928, physicians were finally able to make some real clinical progress, based on the fruits of scientific research.[57]

> Then [in 1937] came the explosive news of sulfanilamide, and the start of the real revolution in medicine.... We knew that other molecular variations of sulfanilamide were on their way from industry, and we heard about the possibility of penicillin and other antibiotics; we became convinced overnight that nothing lay beyond reach for the future.[58]

Alexander Fleming was a Scottish bacteriologist working in London just after the end of the First World War. During the war he had been working on wounds and their resistance to infection and, one night, he accidentally left a Petri dish full of staphylococcus out on the bench while he went away on vacation. When he got back he found that some mold, which had grown in the dish, appeared to have killed off all of the staph around it.[59] After a few experiments, he did not find the result to be clinically promising, yet he nonetheless published a paper on his finding. Ten years later, this result was rediscovered by Howard Florey and Ernst Chain, who then found Fleming's original paper and performed the crucial experiment on mice, isolating penicillin, which saw its first clinical use in 1941.[60]

In his book, *The Rise and Fall of Modern Medicine*, James Le Fanu goes on to list the cornucopia of medical discovery and innovation that followed: cortisone (1949), streptomycin (1950), open heart surgery (1955), the polio vaccine (also 1955), kidney transplantation (1963), and the list goes on.[61] With the development of chemotherapy (1971), in vitro fertilization (1978), and angioplasty (1979), we are a long way from Lewis Thomas's time when the primary job of the physician was to diagnose and simply attend to the patient because nothing much could be done as the illness took its course. Clinical medicine could finally enjoy the benefit of all that basic science.

But it is now time to consider a skeptical question: to what extent can all these clinical discoveries be attributed to science (let alone the scientific attitude)? Le Fanu raises this provocative question by noting that a number of the "definitive" moments in medical history during the twentieth century had little in common. As he notes, "the discovery of penicillin was not the product of scientific reasoning but rather an accident."[62] Yet even if this is true, one needs to be convinced that the other discoveries were not directly attributable to scientific inquiry.

Le Fanu writes, "The paths to scientific discovery are so diverse and depend so much on luck and serendipity that any generalisation necessarily appears suspect."[63] Le Fanu here explores, though he did not invent, the idea that some of the medical breakthroughs of the twentieth century may be thought of not as the direct fruit of scientific research, but instead as "gifts of nature." Selman Waksman, a Nobel Prize winner for medicine for his discovery of streptomycin (and the person who coined the term *antibiotic*)[64] argued—after receiving his prize—that antibiotics were a "purely fortuitous phenomenon." And he was not just being humble. But, as Le Fanu notes, this view was so heretical that many believed it must be wrong.[65]

Can one make a case for the idea that the breakthroughs of modern medicine were due not to "good science" but rather to "good fortune"? This view stretches credibility and, in any case, it is based on the *wrong* view of science. If one views science as a methodological enterprise, where one must follow a certain number of steps in a certain way and scientific discovery comes out at the other end, then perhaps it is arguable whether science is responsible for the discoveries of clinical medicine. Fleming, at least, followed no discernible method. Yet based on the account of science that I am defending in this book, I think it is clear that both the series of breakthroughs in the late nineteenth century and the transition to the fruits of clinical science that started in the early twentieth century were due to the scientific attitude.

For one thing, it is simply too easy to say that penicillin was discovered by accident. While it is true that a number of chance events took place (nine cold days in a row during a London summer, the fact that Fleming's lab was directly above one in which another researcher was working on fungus, that Fleming left a Petri dish out while he went on vacation), this does not mean that just any person who saw what Fleming saw in the Petri dish would have made the discovery. We do not need to attribute the discovery, perhaps, to Fleming's particular genius, but we do not need to attribute it to accident either. No less a giant of medicine than Louis Pasteur once observed that "chance favors the prepared mind." Accident and random events do occur in the lab, but one has to be in the proper mental state to receive them, then probe things a little more deeply, or the benefit is lost. Having the scientific curiosity to learn from empirical evidence (even as the result of an accident), and then *change one's beliefs* on the basis of what one has learned, is what it means to have the scientific attitude. Nature may provide the "fruits," but it is our attitude that allows us to recognize and understand them.

When Fleming saw that there were certain areas in the Petri dish where staphylococcus would not grow—because of (it seemed) contamination from outside spores—he did not simply throw it in the trash and start again. He tried to get to the bottom of things. Even though, as noted, he did not push the idea of clinical applications (for fear that anything powerful enough to kill staph would also kill the patient), he did write a paper on his discovery, which was later discovered by Florey and Chain, who set to work to identify the biochemical mechanisms behind it.[66] Group scrutiny of individual ideas is what led to the discovery.

> Finally, in a classic experiment Chain and Florey demonstrated that penicillin could cure infections in mice: ten mice infected with bacterium streptococcus were divided into two groups, with five to be given penicillin and five to receive a placebo. The "placebo" mice died, the "penicillin" mice survived.[67]

While it is easy to entertain students with the story that penicillin was discovered by accident, it most assuredly was *not* an accident that this discovery was then developed into a powerful drug that was capable of saving millions of lives. That depended on the tenacity and open-mindedness of hundreds of researchers to ask all of the right critical questions and follow through with experiments that could test their ideas. Indeed, one might view the modern era's expectation that the effectiveness of *every* medical treatment should be tested through double-blind randomized clinical trials as one of the most effective practical fruits of those medical researchers who first adopted the scientific attitude. Scientific discovery is born not merely from observing accidents but, even where accidents occur, testing them to see if they hold up.

What changed in medicine during the eighty years (1860–1940) from Pasteur to penicillin? As Porter notes, during this time "one of the ancient dreams of medicine came true. Reliable knowledge was finally attained of what caused major sickness, on the basis of which both preventions and cures were developed."[68] This happened not simply because medical research underwent a scientific revolution. As we have seen, the breakthroughs of Pasteur, Koch, Lister, and others were all either resisted, misunderstood, mishandled, or ignored for far too long to give us confidence that once the knowledge was in place, practitioners would have found it. What more was required were the social forces that led to the transformation of this knowledge into clinical practice that, I have argued, was embedded in a *change in attitude* about how medical education and the practice of medicine should be organized. Once physicians started to think of themselves as a profession rather than a band of individual practitioners, things began to happen. They read one another's work. They scrutinized one another's practices. Breakthroughs and discoveries could still be resisted, but for the first time those who did so faced professional disapproval from their peers and, eventually, the public that they served. As a growing majority of practitioners embraced the scientific attitude, the scrutiny of individual ideas became more common ... and scientific medicine was born.

Conclusion

Within medicine we can see how the employment of an empirical attitude toward evidence, coupled with acceptance of this standard by a group who then used it to critique the work of their peers, was responsible for transforming a field that was previously based on superstition and ideology into a modern science. This provides a good example for the social sciences and other fields that now wish to come forward as sciences. The scientific attitude did not work just for physics and astronomy (and medicine) in the past. It is still working. We can have a modern scientific revolution in previously unscientific fields, if we will just employ the scientific attitude.

Yet we still must face the problem of those who reject the scientific attitude outright. Those who do not seem to understand that ideologically based beliefs like intelligent design are not scientific. Those who indulge in denialism about well-warranted theories like global warming based on a fundamental misunderstanding of how science works. We will deal with those problems in chapter 8. First, however, we must face an even lower low. In the present chapter, we have seen the scientific attitude at its best. In chapter 7 we will explore it at its worst.

7 Science Gone Wrong: Fraud and Other Failures

To anyone who cares about the scientific attitude, one might think that fraud is a perfunctory topic. People who commit fraud are just cheats and liars, who obviously do not embrace the values of science, right? Why bother to examine it any further than that?

But I suggest that we take a more deliberate approach, for the examination of fraud will help us to understand not only what it means by contrast to have a good scientific attitude, it will also help us to take the measure of all those things that fall just short of fraud. If one has an overly simplistic view of fraud, for instance, one might miss the fact that most fraudsters do not see themselves as deliberately trying to falsify the scientific record, but instead feel entitled to a shortcut because they think that the data will ultimately bear them out. This is problematic on many levels, but it is a live issue whether this is a methodological failure or an attitudinal one. By deceiving themselves into thinking that it is all right to cut a few corners in their procedure, does this pave the way for the later commission of fraud, or is this already fraud itself? Actions matter, but so do intentions. If one starts off not intending to falsify anything, but only to shape the data one needs to perform an experiment, at what point do things go off the rails? Might there be a connection between the sorts of sloppy research practices we examined earlier in this book (like p-hacking and cherry picking) and the later falsification or fabrication of data that constitutes fraud itself? As a normative account, the scientific attitude can help us to sort through these problems.

We must start, though, by facing the problem at its worst. Fraud is the intentional fabrication or falsification of the scientific record.[1] In the case of mere error, one's fidelity to science is not at issue, for one can make a mistake without the intent to deceive. But in the case of fraud—where any flaws are deliberate—one's commitment to the scientific attitude is squarely in question. In any activity as open and dependent upon the work of others as science, this is the one thing that cannot be tolerated. When one signs on to be a scientist, one is making a voluntary commitment to be open and honest in learning from experience. By committing fraud, one is putting oneself and one's advancement ahead of this. Ideology, money, ego, and self-interest, are supposed to take a back seat to evidence. It is sometimes said that—because scientific ideas can come from anywhere—there are no heretics in science. But fraud is the one true form of scientific heresy; it is not that one's *theories* are different, it is that these theories are based on invented data. Thus fraud is seen as much worse than error, for fraud is by definition intentional and what is at stake is nothing less than a betrayal of the scientific attitude itself.

Mere error turns out not to be very scary for science. As long as one has the right attitude about learning from empirical evidence, science is well equipped to deal with mistakes. And this is a good thing, because the history of science is replete with them. I am not here talking about the pessimistic-inductivist claim that in the long run most of our scientific beliefs will turn out to be false.[2] I am talking about the enormous errors and dead ends that knocked science off its track for centuries at a time. Phlogiston. Caloric. Ether. Yet it is important to point out that these mistakes were not frauds, and, in fact, they were in some cases pivotal for coming up with better scientific theories. Without phlogiston we might not have discovered oxygen. Without caloric we might not understand thermodynamics. Why is this? Because science is expected to learn from error. If one embraces the scientific attitude and follows the evidence, error will eventually be rooted out. I suppose one might try to make the same case for errors that are introduced by fraud—for if science is self-correcting these too will eventually be found and fixed. But it is just such an enormous waste of time and resources to chase down the results of *intentional* error that this is where most scientists draw the line. It is not just the *consequences* of fraud that lead it to be so despised, it is the break of faith with scientific values. Nature is subtle enough; scientists don't care to deal with any additional challenges created by deception.

But it is important to remember that there is another source of defect in scientific work. Between fraud and honest error, there is a murky category where it is not entirely clear whether one's motives are pure. As we saw in chapter 5, scientific error can come from fraud, but it can also come from sloppiness, cognitive bias, willful ignorance, or laziness. I hope already to have established that the scientific attitude is a robust tool to mitigate error, whatever the source. Yet here—on the verge of claiming that fraud is the worst sort of affront one can make against the scientific attitude—we should revisit the question of how to divide these sources of error along the lines of intentional versus unintentional motivations.

The key here is to be explicit in the definition of fraud. If we define fraud as the *intentional* fabrication or falsification of scientific data, then there are two possible ways to read this:

(1) If one has committed fraud, then one has intentionally fabricated or falsified data.

(2) If one has intentionally fabricated or falsified data, then one has committed fraud.

As we know from logic, these two statements do not imply one another and it is therefore possible for one of them to be true, while the other is not. In this case, however, I think that both of them are true. If one is committing fraud, then it has to be intentional. As we saw in the broader definition of "research misconduct" (given in note 1 to this chapter), if a mistake is due to "honest error" or "difference of opinion," it is not considered fraud. For fraud, it is necessary to have the intention to deceive. But we must then ask whether if one commits fabrication or falsification intentionally, this is *sufficient* for fraud. It is. Fabrication and falsification are not just any kinds of errors; by their very definition, they cannot be done by accident. So the minute one engages in these kinds of behaviors, it seems to automatically constitute fraud. We might thus seek to define fraud by combining (1) and (2) into the following biconditional statement: "One commits fraud if and only if one intentionally fabricates or falsifies scientific data."[3] Yet this still leaves open the crucial question of how to define intentionality. Is there perhaps a better way to characterize fraud to make this clear?

Let's turn now to the scientific attitude, and see what leverage this might give us in understanding the concept of scientific fraud. Throughout this book, I have been arguing that the scientific attitude is what defines science; that it can help us to understand what is special about science and why there is unique warrant behind scientific beliefs. Since I just got done saying that fraud is the worst sort of crime one can commit against science, it would seem to follow that fraud must constitute a complete repudiation of the scientific attitude. But now consider the following two ways of interpreting this claim:

(3) If one has committed fraud, then one does not have the scientific attitude.

(4) If one does not have the scientific attitude, then one has committed fraud.

There is an obvious problem here, for I think that thesis (3) is true, but (4) is not. How could that be? With thesis (3) it seems obvious that if someone has committed fraud, that person does not have the scientific attitude. To fabricate or falsify data is in direct conflict with the idea that one cares about empirical evidence and is committed to holding and changing one's beliefs on this basis. So why then is thesis (4) false? The issue is a subtle one, for in some situations thesis (4) may well be true, but the issue here is that it is not *necessarily* true in all cases.[4] To say that "if one does not have the scientific attitude, then one has committed fraud" is to make a large presumption. First, one has to be investigating in an empirical field; literature does not have the scientific attitude, but so what? Second, one is presuming that if one has the wrong attitude during empirical inquiry, one will definitely act on it. But we know from human behavior that this is not always the case. And third, what about the issue of intentionality? From thesis (2) above, it seems that if we have committed an intentional error then we have committed fraud. But the issue here is that there are many different levels of intentionality and many different reasons why someone might not have the scientific attitude.

As we saw in chapter 5, it could be that some researchers are the victims of unconscious cognitive bias. Or perhaps they are just lazy or sloppy. Are they also frauds? There could be a whole host of subterranean psychological reasons why not having the scientific attitude is not someone's fault. It is not necessarily intentional when someone violates the scientific attitude. But here is the key question. What about those cases in which someone *does* intentionally engage in shady research practices? What about all of those less-than-above-board research practices like p-hacking or cherry picking data that I was railing against in chapter 5? Why aren't *those* considered fraud the minute they are done intentionally? But the relevant question to making a determination of fraud is not just whether those actions are done intentionally, it is whether they also involve fabrication or falsification. Remember our working definition of fraud: the intentional fabrication or falsification of scientific data. (Recall too that this is a biconditional relationship.) The reason that p-hacking isn't normally considered fraud isn't that the person who did it didn't mean to; it's that, as egregious as it may seem, p-hacking is not quite up to the level of falsifying or fabricating data. One is *misleading* one's scientific colleagues, perhaps, but not fabricating evidence. One may be leaving a study open to get more data so that one can publish, but this is not quite falsifying.[5]

Consider an analogy with lying. To tell a bald-faced lie is to say something false while knowing that it is false. But what about those instances where we have not lied, but we have not exactly told the whole truth either? This is patently dishonest, but not (quite) the same as lying. This is precisely the analogy we are looking for to mark off the difference between questionable research practices and fraud. P-hacking, selective data reporting, and the like are not considered fraud by the standard definition because they do not involve fabrication or falsification of data. Yet they are not altogether honest either.[6] They are intentional deceptions that fall short of fraud. They may be a crime against the scientific attitude, but they are not quite a felony. If done intentionally we should hope that these practices can be exposed and discouraged—and even that the scientific attitude (which helps us to understand what is so wrong about fraud) may provide a tool to help us do this—but this does not mean that we should confuse them with fraud.

The scientific attitude may be thought of as a spectrum, with complete integrity at one end and fraud at the other. The criterion that delineates fraud is the intentional fabrication or falsification of data. One can fall short of this either because one has made an unintentional mistake or because one was misleading but it did not rise to the level of fabrication or falsification. Into the latter class, I would put many of those "misdemeanors" against the "degrees of freedom" one has as a scientific researcher.[7] The scientific attitude is a matter of degree; it is not all or nothing. Fraud may be thought of as occurring when someone violates the scientific attitude *and* their behavior rises to the level of fabrication or falsification. Yet a researcher can have an eroded sense of the scientific attitude and not go this far. (Cheating on the scientific attitude thus seems necessary, but not sufficient, for committing fraud.)

While it seems valuable to draw a bright line where one crosses over into fraud, this does not mean that "anything goes" short of it. Using the scientific attitude to mark off what is special about science should be able to help us with both tasks. In this chapter, I will argue that we may use the scientific attitude to gain a better understanding of what is so egregious about fraud and to police the line between fraud and other failures of the scientific attitude. In doing so, I hope to illuminate the many benefits that an understanding of the scientific attitude may offer for identifying and discouraging those shoddy research practices that fall just short of fraud as well. As we saw in chapter 5, the scientific attitude can help us to identify and fight all manner of error. But the proper way to do this is to understand each error for what it is. Some will find anything short of complete honesty in the practice of science to be deplorable. I commend this commitment to the scientific attitude. Yet science must survive even when some of its practitioners—for whatever reason—occasionally misbehave.

Why Do People Commit Fraud?

The stereotypical picture of the scientific fraudster as someone who just makes up data is not necessarily accurate. Of course this does occur, and it is a particularly egregious form of fraud, but it is not the only or even the most common kind. Just as guilty are those who think that they already know the answer to some empirical question, and can't be bothered to take the time—due to various pressures—to get the data right.

In his excellent book *On Fact and Fraud,* David Goodstein provides a bracing analysis of numerous examples of scientific fraud, written by someone who for years has been charged with investigating it.[8] After making the customary claim that science is self-correcting, and that the process of science will eventually detect the insertion of any falsehood (no matter whether it was intentional or unintentional),[9] Goodstein goes on to make an enormously provocative claim.[10] He says that in his experience most of those who have committed fraud are not those who are deliberately trying to insert a *falsehood* into the corpus of science, but rather those who have decided to "help things along" by taking a shortcut to some *truth* that they "knew" would be vindicated.[11] This assessment should at least give us pause to reconsider the stereotypical view of scientific fraud.[12] Although there are surely examples of fraud that have been committed by those who deliberately insert falsehoods into the corpus of science, what should we say about the "helpers"? Perhaps here the analogy with the liar (who intentionally puts forth a falsehood) is less apt than that of the impatient egoist, who has the hubris to short-circuit the process that everyone else has to follow.

Yet, seen in this light, scientific fraud is a matter not merely of bad motives, but of having the arrogance to think that one deserves to take a shortcut in how science is done.

It is notable that concern with hubris in the search for knowledge predates science. In his dialogues, Plato makes the case (through Socrates) that false belief is a greater threat to the search for truth than mere error.[13] Time and again, Socrates exposes the ignorance of someone like Meno or Euthyphro who *thought* they knew something, only to find out quite literally that they didn't know what they were talking about. Why is this important? Not because Socrates feels that he himself has all the answers; Socrates customarily professes ignorance. Instead, the lesson seems to be that error is easier to recover from than false belief. If we make an honest mistake, we can be corrected by others. If we accept that we are ignorant, perhaps we will go on learning. But when we think that we already know the truth (which is a mindset that may tempt us to cut corners in our empirical work) we may miss the truth. Although the scientific attitude remains a powerful weapon, hubris is an enemy that should not be underestimated. Deep humility and respect for one's own ignorance is at the heart of the scientific attitude. When we violate this, we may already be on the road to fraud.[14]

If some, at least, commit fraud with the conviction that they are merely hurrying things along the road to truth, is their *attitude* vindicated? No. Just as we would not vindicate the vigilante who killed in the name of justice, the "facilitator of truth" who takes shortcuts is guilty not just of bad actions but of bad intent. Even with so-called well-intentioned fraud, the deceit was still intentional. One is being dishonest not merely in one's actions but in one's mind. Fraud is the *intentional* fabrication or falsification of evidence, in order to convince someone else to believe what we want them to believe. But without the most rigorous methods of gathering this evidence, there is no warrant. Merely to be right, without justification, is not knowledge. As Socrates puts it in *Meno*, "right opinion is a different thing than knowledge."[15] Knowledge is *justified* true belief. Thus fraud short-circuits the process by which scientists formulate their beliefs, even if one guesses right. Whatever the motive, one who commits fraud is doing it with full knowledge that this is not the way that science is supposed to be done. Whether one thought that one was "inserting a falsehood" or "helping truth along" does not matter. The hubris of "thinking that you are right" is enough to falsify not just the result but the process. And in a process as filled with serendipity and surprise as science, the danger of false belief is all around us.

The Thin Crimson Line

One problem with judging fraud is the use of euphemistic language in discussing it. Understanding the stakes for one's academic career, universities are sometimes reluctant to use the words "fraud" or "plagiarism" even in cases that are quite clear-cut.[16] If someone is found guilty (or sometimes even suspected) of fraud, they are all but "excommunicated" from the community of scientists. Their reputation is dishonored. Everything they have ever done—whether it was fraudulent or not—will be questioned. Their colleagues and coauthors will shun them. Sometimes, if federal money is mismanaged or they live in a country with strict laws, they may even go to jail.[17] Yet the professional judgment of one's peers is often worse (or at least more certain) than any criminal punishment. Once the taint of fraud is in the air, it is very hard to overcome.[18] It is customary for someone who has been found guilty of fraud simply to leave the profession.

One recent example of equivocating in the face of fraud is the case of Marc Hauser, former Professor of Psychology at Harvard University, who was investigated both by Harvard and by the Office of Research Integrity (ORI) at the National Institutes of Health. The results of Harvard's own internal investigation were never made public. But in the federal finding that came out some time later, the ORI found that half of the data in one of Hauser's graphs was fabricated. In another paper he "falsified the coding" of some data. In another he "falsely described the methodology used to code the results for experiments." And the list goes on. If this isn't fraud, what is? Yet the university allowed Hauser to say—before the federal findings came out—that his mistakes were the result of a "heavy workload" and that he was nonetheless willing to accept responsibility "whether or not I was directly involved." At first Hauser merely took a leave of absence, but after his faculty colleagues voted to bar him from teaching, he quietly resigned. Hauser later worked at a program for at-risk youth.[19]

Although many may be tempted to use the term "research misconduct" as a catch-all phrase that includes fraud (or is a euphemism for fraud), this blurs the line between intentional and unintentional deception. Does research misconduct also include sloppy or careless research practices? Are data fabrication and falsification in the same boat as improper data storage? The problem is a real one. If a university is trying to come up with a policy on fraud it might write somewhat differently than if it were writing a policy on scientific misconduct. As Goodstein demonstrates in his book, the latter can tempt us to include language about nonstandard research practices as something we may want to discourage and even punish but does not rise to the level of fraud. Goodstein writes, "There are many practices that are not commonly accepted within the scientific community, but don't, or shouldn't, amount to scientific fraud."[20] What difference does this make? Some might argue that it doesn't matter at all. That even bad research practices like "poor data storage or retention," "failure to report discrepant data," or "overinterpretation of data" represent a failure of the scientific attitude. As previously noted, the scientific attitude isn't all or nothing. Isn't engaging in "deviation from accepted practices"[21] also to be discouraged? Maybe so, but I would argue that there is a high cost for not differentiating this from fraud.

Without a sharp line, it may sometimes be difficult even for researchers themselves to tell when they are on the verge of fraud. Consider again the example of cold fusion. Was this deception or self-deception—and can these be cleanly separated?[22] In his book *Voodoo Science*, Robert Park argues that self-delusion evolves imperceptibly into fraud.[23] Most would disagree, because fraud is intentional. As Goodstein remarks, self-delusion and other examples of human foibles should not be thought of as fraud.

> Mistaken interpretations of how nature behaves do not and never will constitute scientific misconduct. They certainly tell us something about the ways in which scientists may fall victim to self-delusion, misperceptions, unrealistic expectations, and flawed experimentation, to name but a few shortcomings. But these are examples of all-too-human foibles, not instances of fraud.[24]

Perhaps we need another category. Goodstein argues that even though the cold fusion case was not fraud it comes close to what Irving Langmuir calls "pathological science," which is when "the person always thinks he or she is doing the right thing, but is led into folly by self-delusion."[25] So perhaps Park and Goodstein are both right: even if self-delusion is *not* fraud, it may be a step on the road that leads there. I think we need to take seriously the idea that what starts as self-delusion might later (like hubris) lead us into fraud. The question here is whether tolerating or indulging in self-delusion for long enough erodes our attitude toward what good science is supposed to look like.[26]

Moreover, even if we are solely concerned (as we are now) with intentional deception, it might be a good idea to examine any path that may lead there. It is important to recognize that self-delusion, cognitive bias, sloppy research practices, and pathological science are all dangerous—even if we do not think that they *constitute* fraud—precisely because if left unchecked they might erode respect for the scientific attitude, which can *lead* to fraud. But this does not mean that actual fraud should not be held distinct. Neither should there be any excuse for lack of clarity in university policies over what actually *is* fraud, versus what practices we merely wish to discourage. We are right to want to encourage researchers to have impeccable attitudes about learning from evidence, even if we must also draw a line between those who are engaging in questionable or nonstandard research practices and those who have committed fraud.

Any lack of clarity—or lack of commitment actually to *use* the word "fraud" in cases that are unequivocal—can be a problem, for it allows those who have committed fraud sometimes to hide behind unspecified admissions of wrongdoing and cop to the fact that they made mistakes without truly accepting responsibility for them. This does a disservice not only to the majority of honest scientists, but also to those who have not (*quite*) committed fraud, for it makes the community of scientists even more suspicious when someone has committed only a mistake (e.g., faulty data storage) yet has not committed full-blown fraud.[27] If fraud is defined merely as one type of scientific misconduct, or we use the latter phrase as a euphemism for the former, whom does this serve?

If the scientific attitude is our criterion, when we find fraud we should name and expose it. This will act as a deterrent to others and a signal of integrity for science as a whole.

> We must be vigilant to find and expose such wrongdoers, careful at the same time not to spread the blame beyond where it belongs and unintentionally stifle the freedom to question and explore that has always characterized scientific progress.[28]

When an allegation of fraud is made public, the punishment from one's community can (and should) be swift and sure. But first it must not be covered up. We can acknowledge the pressures to do so, but these can tarnish the reputation of science. For when blame is not cast precisely where it should be—and some suspect that fraud is too often excused or covered up—the unintended consequence can be that an injustice is done to those who are merely *accused* of it. When fraud is selectively punished, those who are only accused may be presumed guilty. We see evidence of this in the previously mentioned scandals over reproducibility and article retraction. Scientific errors sometimes happen. Some studies are irreproducible and/or can be retracted for reasons that have nothing whatsoever to do with fraud. Yet if there is no sharp line for what constitutes fraud—and we retreat into the weasel words "research misconduct"—it is far too easy to say "a pox on all your houses" and look only at external events (like article retraction) and assume that this is a proxy for bad intent. The sum result is that some fraudsters are allowed to get away with it, while some who have not committed fraud are falsely accused. None of this is good for science.

When left to scientists rather than administrators, there is usually no equivocating about naming and punishing actual instances of fraud. Indeed, I see it as one of the virtues of using the scientific attitude to distinguish science from nonscience that it explains why scientists are so hard on fraud. If we talked more about the scientific attitude as an essential feature of science, this might allow scientists more easily to police the line between good and bad science.[29] Some may wonder why this would be. After all, if the process of group scrutiny of individual ideas in science is so good, it will catch all types of errors, whether they were committed intentionally or not. But this misses the point, which is that science is precisely the kind of enterprise where we must count on most policing to be self-policing. If science were a dishonest enterprise where everyone cheated—and it was the job of peer reviewers to catch them—science would break down. Thus fraud is and should be recognized as different in kind from other scientific mistakes, for it represents a breach of faith in the values that bind scientists together.

The Vaccine–Autism Debacle

We are now in a position to consider the impact that scientific fraud can have not just on scientists but on the entire community of people who rely on science to make decisions about their daily lives. In 1998, Dr. Andrew Wakefield published a paper with twelve coauthors in the prestigious British medical journal *Lancet*, which claimed to have found a link between the classic MMR triple vaccine and the onset of autism. If true, this would have been an enormous breakthrough in autism research. Both the public and the press demanded more information, so along with a few of his coauthors Wakefield held a press conference. Already, questions were being raised about the validity of the research. As it turned out, the paper was based on an extremely small sample of only twelve children. There were, moreover, no controls; all of the children in the study had been vaccinated and had autism. While this may sound to the layperson like good evidence of a causal link, to someone with training in statistics, questions will naturally arise. For one, how did the patients come to the study? This is important: far from being a randomized double-blind clinical study (where researchers randomly test their hypothesis on only half of a sample population, with neither the subject nor the researcher knowing who is in which half), or even a "case control study" (where investigators examine a group that has been naturally exposed to the phenomenon in question),[30] Wakefield's paper was a simple "case series" study, which is perhaps the equivalent of finding out by accident that several people have the same birthday, then mining them for further correlations. Obviously, with the latter, there can be a danger of selection bias. Finally, a good deal of the study's evidence for a correlation between vaccines and autism was based on a short timeline between vaccination and onset of symptoms, yet this was measured through parental recollection.

Any one of these things would be enough to raise suspicions in the minds of other researchers, and they did. For the next several years, medical researchers from all over the world performed multiple studies to see if they could recreate Wakefield's proposed link between vaccines and autism. A good deal of speculation focused on the question of whether thimerosal in the MMR shot might have caused mercury poisoning. In the meantime, just to be safe, several countries stopped using thimerosal while research was underway. But, in the end, none of the studies found any link.

> Epidemiologists in Finland pored over the medical records of more than two million children... finding no evidence that the [MMR] vaccine caused autism. In addition, several countries removed thimerosal from vaccines before the United States. Studies in virtually all of them—Denmark, Canada, Sweden, and the United Kingdom—found that the number of children diagnosed with autism continued to rise throughout the 1990s, after thimerosal had been removed. All told, ten separate studies failed to find a link between MMR and autism; six other groups failed to find a link between thimerosal and autism.[31]

Meanwhile, a few stunning facts about Wakefield's original study came to light. In 2004, it was discovered that Wakefield had been on the payroll of an attorney who was planning a massive lawsuit against an MMR vaccine manufacturer. Worse, it turned out that almost half the children who had been reported on in Wakefield's study had been funneled to him through the lawyer. Finally, it was learned that just before Wakefield published his study, he had filed a patent for a competing vaccine to the classic MMR shot.[32] Far from mere selection bias, this was a massive undisclosed conflict of interest that raised numerous questions over Wakefield's motives. Within days, ten of Wakefield's coauthors took their names off the study.

But by this point it was too late. The public had already heard the rumors and vaccination rates had begun to drop. In Ashland, Oregon, there was a 30 percent vaccination exemption rate. In Marin County, California, the exemption rate was more than three times the rest of the state.[33] With such pockets of vaccine resistance, doctors began to worry about "herd immunity," which is when the vaccination rate falls so low that one can no longer count on the "free rider" benefit of remaining unvaccinated in a community where most others have been vaccinated. And the results were devastating. After being beaten to a standstill, measles, whooping cough, diphtheria, and other diseases began to make a comeback:

> [Measles] is the most infectious microbe known to man and has killed more children than any other disease in history. A decade after the World Health Organization (WHO) declared the virus effectively eradicated everywhere in the Americas save for the Dominican Republic and Haiti, declining vaccination rates have led to an explosion of outbreaks around the world. In Great Britain, there's been more than a thousandfold increase in measles cases since 2000. In the United States, there have been outbreaks in many of the country's most populous states, including Illinois, New York, and Wisconsin.[34]

It didn't help that many in the media were whipping up the story, trying to tell "both sides" of the vaccine "controversy."[35] Meanwhile, many parents of autistic children didn't care about any alleged irregularities in Wakefield's work. He continued to speak at autism conferences worldwide, where he was treated as a hero. When the *Lancet* finally retracted his paper (in 2010), and Wakefield was stripped of his medical license in Britain, conspiracy theories began to run wild. Why was his work being suppressed? Angry parents (including a number of Hollywood celebrities) were already organized and furious with what they saw as a cover up. If thimerosal wasn't dangerous, why had it been removed?

Then in 2011, definitive word came: Wakefield's work was a fraud. In addition to the severe conflict of interest noted above, Brian Deer (an investigative journalist who had already broken a good deal of the earlier revelations in 2004) finally had a chance to interview the parents of Wakefield's patients and examine their medical records. And what he found was shocking. "No case was free of misreporting or alteration."[36] Wakefield had altered the medical records of every single child in the study.

> Three of nine children reported with regressive autism did not have autism diagnosed at all. Only one child clearly had regressive autism.
>
> Despite the paper claiming that all 12 children were "previously normal," five had documented pre-existing developmental concerns.
>
> Some children were reported to have experienced first behavioural symptoms within days of MMR, but the records documented these as starting some months after vaccination. …
>
> The parents of eight children were reported as blaming MMR, but 11 families made this allegation at the hospital. The exclusion of three allegations—all giving times to onset of problems in months—helped to create the appearance of a 14 day temporal link.
>
> Patients were recruited through anti-MMR campaigners, and the study was commissioned and funded for planned litigation.[37]

The *British Medical Journal* (perhaps the second-most prestigious medical journal in Britain, after the *Lancet*) took the unprecedented step of accepting Deer's work as definitive evidence of fraud and, after it had been peer reviewed, published his paper alongside their own editorial, which concluded that "clear evidence of falsification of data should now close the door on this damaging vaccine scare" and called Wakefield's work an "elaborate fraud."[38] They concluded:

> Who perpetrated this fraud? There is no doubt that it was Wakefield. Is it possible that he was wrong, but not dishonest: that he was so incompetent that he was unable to fairly describe the project, or to report even one of the 12 children's cases accurately? No. A great deal of thought and effort must have gone into drafting the paper to achieve the results he wanted: the discrepancies all led in one direction; misreporting was gross.[39]

A few months later another commentator called Wakefield's fraud "the most damaging medical hoax of the last 100 years."[40] Four year later, in early 2015, there was a measles outbreak with over a hundred confirmed cases across fourteen states in the US.[41]

As we can see, scientific fraud is ugly and the fallout can be massive.[42] Yet one of the most interesting parts of the story is the enormous scorn of

Wakefield's work by the scientific community (juxtaposed, unfortunately, against public confusion and willful ignorance enabled by the media), *before* he was proven to be a fraud. Why did this occur? If fraud must be intentional, how did the scientific community seem to reach a consensus in advance of seeing proof that Wakefield had manipulated data? The answer is that although fraud is perhaps the most egregious form of intentional misconduct, it is not the only kind of cheating one can do. Once it had come to light that Wakefield had an enormous undisclosed conflict of interest, his intentions were suspicious. Even though no one had yet proven that his financial interests had colored his scientific work, where there was so much smoke, few in the scientific community could see how there was not a huge fire behind it. Since Wakefield had already forsaken a core principle of scientific practice—that one must disclose in advance all possible conflicts of interest—many concluded that he didn't deserve the benefit of the doubt. And they were right. Yet one mourns that the scientific community's self-correction in this case still has not made its way to all corners of the general population.[43]

On a Happier Note

I would like to end on a brighter note. In this chapter, we have encountered perhaps the ugliest face of science. But what should a scientist do if his or her theory isn't working? Where the time and career pressures are massive and the data are coming out all wrong?

A few years before Andrew Wakefield's paper, a little-known British astronomer named Andrew Lyne stood before several hundred colleagues at the American Astronomical Society meeting in Atlanta, Georgia. He had been invited to give a paper on his stunning discovery of a planet orbiting a pulsar. How could that happen? A pulsar is the result of a star that has exploded in a supernova, which theoretically should have destroyed anything even close to its orbit. And yet, after rechecking his results, the planet remained, so Lyne published his paper in the prestigious journal *Nature*. But now there was a problem. A few weeks before his trip to Atlanta, Lyne discovered a crucial error in one of his calculations: he had forgotten to account for the fact that the Earth's orbit was elliptical rather than circular. This was a mistake from first-year physics. When he made the correction, "the planet disappeared." But, standing there that day in front of his colleagues, Lyne made no excuse for himself. He told the audience what he had found and then told them why he had been wrong, after which they gave him a standing ovation. It was "the most honorable thing I've ever seen," said one astronomer who was present. "A good scientist is ruthlessly honest with him- or herself, and that's what you've just witnessed."[44]

That is the true spirit of the scientific attitude.

8 Science Gone Sideways: Denialists, Pseudoscientists, and Other Charlatans

We turn now from fraud—which is a matter of accepting the standards of science, then intentionally violating them—to the case of denialists and pseudoscientists, who may misunderstand or not care at all about the standards of scientific evidence or, to the extent that they do, not enough to modify or abandon their ideological beliefs.

Many scientists have found it incredible in recent years that their conclusions about empirical topics are being questioned by those who feel free to disagree with them based on nothing more than gut instinct and ideology. This is irrational and dangerous. Denialism about evolution, climate change, and vaccines has been stirred up in recent years by those who have an economic, religious, or political interest in contradicting certain scientific findings. Rather than merely wishing that particular scientific results weren't true, these groups have resorted to a public relations campaign that has made great strides in undermining the public's understanding of and respect for science. In part, this strategy has consisted of attempts to "challenge the science" by funding and promoting questionable research—which is almost never subject to peer review—in order to flood news outlets with the appearance of scientific controversy where there is none. The result has been a dangerously successful effort to subvert the credibility of science.

As we saw in the last chapter, the scientific attitude is sometimes betrayed even by scientists. Yet a proportionally greater threat may arise from those who are outside science: those who either willfully or unwittingly misunderstand the process of science, who are prepared to deny scientific results that do not jibe with their ideological beliefs, who are only pretending to do science in order to further their pet theories, who cynically exploit others' ignorance while they profit, or who fool themselves by being overly gullible. It is important to realize, however, that these errors can be committed both by those who are lying to others (who are doing false science or are rejecting science falsely) and by those who are being lied to (who have not bothered to educate themselves in the skills necessary to form well-warranted beliefs). Whether conscious of their betrayal of good scientific principles or not, in an age where climate change deniers routinely use Facebook and Twitter to spread their benighted ideology, and intelligent design theorists have a website to share their cherry-picked doubts about evolution, we are all responsible for any distance between our reliance on the fruits of science and our sometimes woefully misinformed conception of how scientific beliefs are formed.

The two most pernicious forms of cheating on scientific principles by those who are outside science are *denialism* and *pseudoscience*. Although both will be discussed at greater length later in this chapter, let me now define denialism as the refusal to believe in well-warranted scientific theories even when the evidence is overwhelming.[1] The most common reason for this is when a scientific theory conflicts with someone's ideological beliefs (for instance, that climate change is a hoax cooked up by liberals), so they refuse to look at any distasteful evidence. Pseudoscience, by contrast, is when someone seeks the mantle of science to promote a fringe theory about an empirical matter (such as intelligent design), but refuses to change their beliefs even in the face of refutatory evidence or methodological criticism by those who do not already believe in their theory. As we will see, it is difficult to draw a clear line between these practices because they so often overlap in their tactics, but both are united in their repudiation of the scientific attitude.

Everyone has a stake in the justification of science. If our beliefs are being manipulated by those who are seeking to deceive us—especially given that virtually all of us are prewired with cognitive biases that can lead to a slippery slope of gullibility and self-deception—the consequences for scientific credibility are enormous. We may feel justified in believing what we want to believe about empirical matters (falsely judging that if scientists are still investigating there must be a lack of consensus), but if we do this then who do we have to blame but ourselves if the planet is nearly uninhabitable in fifty years? Of course, this is to oversimplify an enormously complex set of psychological circumstances, for there are many shades of awareness, bias, intentionality, and motivation, all of which bear on the nature of belief. As Robert Trivers masterfully demonstrates in his previously cited book *The Folly of Fools*, the line between deception and self-deception may be thin.

Just as scientific researchers may sometimes engage in pathological science due to their own delusions, those who engage in denialism or pseudoscience may believe that they are actually living up to the highest standards of the scientific attitude.

But they are not.[2] And neither are those who uncritically accept these dogmas, never bothering to peek over the wall of willful ignorance at the results of good science. In this chapter, I will explore the mistakes of both the liars and those who are overly credulous. For as I've said, in a day and age where scientific results are at our fingertips, we all bear some responsibility for the warrant behind our empirical beliefs. And, for science, it is a problem either way. Whether someone has lit the fire of denialism about climate change or is merely stopping by to warm their hands, it is still repudiation of a core value of science.[3]

In the last chapter, we explored what happens when a scientific researcher cheats on the scientific attitude. In this chapter, we will consider what happens when those who are *not* doing science—whatever their motive—peddle their convictions to the larger community and cast doubt on the credibility of well-warranted scientific beliefs.

Ideology and Willful Ignorance

Scientists presumably have a commitment to the scientific attitude, which will influence how they formulate and change their beliefs based on empirical evidence. But what about everyone else? Fortunately, many people have respect for science. Even if they do not do science themselves, it is fair to say that most people have respect for the idea that scientific beliefs are especially credible and useful because of the way they have been vetted.[4] For others, their primary allegiance is to some sort of ideology. Their beliefs on empirical subjects seem based on fit not with the evidence but rather with their political, religious, or other ideological convictions. When these conflict—and the conclusions of science tread on some sacred topic on which people think that they already know the answer (e.g., whether prayer speeds up healing, whether ESP is possible), this can result in rejection of the scientific attitude.

There has always been a human tendency to believe what we want to believe. Superstition and willful ignorance are not new to the human condition. What *is* new in recent years is the extent to which people can find a ready supply of "evidence" to support their conspiracy-theory-based, pseudoscientific, denialist, or other outright irrational beliefs in a community of like-minded people on the Internet. The effect that group support can have in hardening one's (false) convictions has been well known to social psychology for over sixty years.[5] In these days of 24/7 partisan cable "news" coverage, not to mention Facebook groups, chat rooms, and personal news feeds, it is increasingly possible for those who wish to do so to live in an "information silo," where they are seldom confronted by inconvenient facts that conflict with their favored beliefs. In this era of "fake news" it is possible for people not only to avoid views that conflict with their own, but almost to live in an alternative reality, where their preferred views are positively reinforced and opposing views are undermined. Thus political and religious ideologies—even where they tread on empirical matters—are increasingly "fact free" and reflect a stubborn desire to shape reality toward them.

To say that this is a dangerous development for science would be an understatement. In fact, I think it is so dangerous that I wrote an entire book—*Respecting Truth: Willful Ignorance in the Internet Age*—on the topic of how these forces have conspired to create an increasing lack of respect for the concept of truth in recent years.[6] I will not repeat those arguments here, but I would like to trace out their implications for the debate about the distinctiveness of science.

One important topic here is the role of group consensus. We have already seen that in science, consensus is reached only after rigorous vetting and comparison with the evidence. Community scrutiny plays a key role in rooting out individual error. In science, we look to groups not for reinforcement of our preexisting beliefs, but for criticism. With ideological commitments, however, one often finds little appetite for this and people instead turn to groups for agreement.[7] Yet this feeds directly into the problem of "confirmation bias" (which we've seen is one of the most virulent forms of cognitive bias, where we seek out evidence that confirms our beliefs rather than contradict them). If one wants to find support for a falsehood, it is easier than ever to do so. Thus, in contrast to the way that scientists use groups as a check against error, charlatans use them to reinforce prejudice.

Sagan's Matrix

In his influential book *The Demon-Haunted World: Science as a Candle in the Dark*,[8] Carl Sagan makes the observation that science can be set apart from pseudoscience and other chicanery by two simple principles: openness and skepticism. As Sagan puts it:

> The heart of science is an essential balance between two seemingly contradictory attitudes—an openness to new ideas, no matter how bizarre or counter-intuitive, and the most ruthless skeptical scrutiny of all ideas, old and new.[9]

By "new ideas," Sagan means that scientists must not be closed to the power of challenges to their old way of thinking. If scientists are required to base their beliefs on evidence, then they must be open to the possibility that new evidence may change their minds. But, as he goes on to observe, one must not be so open to new ideas that there is no filter. Scientists cannot be gullible and must recognize that "the vast majority of ideas are simply wrong."[10] Thus these two principles must be embraced simultaneously even while they are in tension. With "experiment as the arbiter," a good scientist is both open and skeptical. Through the critical process, we can sort the wheat from the chaff. As Sagan concludes, "some ideas really are better than others."[11]

Some may complain that this account is too simple, and doubtless it is, but I think it captures an essential idea behind the success of science. Yet perhaps the best measure of the depth of Sagan's insight is to examine its implication for those areas of inquiry that are *not* open or *not* skeptical. Let us now dig a little deeper into denialism and pseudoscience. Although he does not discuss denialism per se, Sagan discusses pseudoscience at length, allowing for an intriguing comparison. What is the difference between denialism and pseudoscience?

Sagan says that pseudoscientists are gullible, and I think that most scientists would be hard pressed to disagree.[12] If one goes in for crystal healing, astrology, levitation, ESP, dowsing, telekinesis, palmistry, faith healing, and the like,[13] one will find little support from most scientists. Yet virtually all of these belief systems make some pretense of scientific credibility through seeking evidential support. What is the problem? It is not that they are not "open to new ideas," but that in some ways they are "too open."[14] One should not believe something without applying a consistent standard of evidence. Cherry picking a few favorable facts and ignoring others is not good scientific practice. Here Sagan cites favorably the work of CSICOP (Committee for the Scientific Investigation of Claims of the Paranormal—now the Committee for Skeptical Inquiry), the professional skeptical society that investigates "extraordinary" beliefs. If science really is open, such claims deserve a hearing.[15] But the problem is that in virtually every case in which real skeptics have investigated such extraordinary claims, the evidence has not held up.[16] They are revealed as pseudoscientific not because they are new or fantastical, but because they are believed without sufficient evidence.

This way of looking at pseudoscience allows for a fascinating contrast with denialism. Although, as noted, Sagan does not make this comparison, one wonders whether he might approve of the idea that the problem with denialism is not that it is not skeptical enough, but that it is insufficiently open to new ideas.[17] When you're closed to new ideas—most especially to any evidence that might challenge your ideological beliefs—you are not being scientific. As Sagan writes, "If you're only skeptical, then no new ideas make it through to you. You never learn anything."[18] Although one desires a much more developed sense of scientific skepticism than Sagan offers here (which I will soon provide), his notion does at least suggest what may be wrong with denialism. The scientific attitude demands that evidence counts because it might change our thinking. But for denialists, no evidence *ever* seems sufficient for them to change their minds. Through embracing the scientific attitude, science has a mechanism for recovering from its mistakes; denialism does not.

So are pseudoscience and denialism more similar or different? I would argue that they have some similarities (and that their demographics surely overlap) but that it is enlightening to pursue the question of their purported differences. Later in this chapter, I will explore these notions in more depth but for now, as a starting point, let's offer a 2 × 2 matrix of what might be regarded as a Sagan-esque first cut on the issue.[19]

	Skeptical	Gullible
Open	Science	Pseudoscience
Closed	Denialism	Conspiracy Theories

Notice that in the extra box I offer the possibility of conspiracy theories. These seem both closed and gullible. How is that possible? Consider the example of someone who argues that NASA faked the Moon landing. Is this a closed theory? It would seem so. No evidence provided by Moon rocks, videotape, or any other source is going to be enough to convince disbelievers. This, it should be noted, is not true skepticism but only a kind of selective willingness to consider evidence that fits with one's hypothesis. Evidence counts, but only relative to one's prior beliefs. What about the claim that Moon-landing deniers are gullible? Here the "skeptical" standard does not apply at all. Anyone who thinks that the US government is capable of covering up something as enormous as a faked Moon landing must either be extremely gullible or have a faith in governmental competence that belies all recent experience. Here the problem is that one's beliefs are not subject to sufficient scrutiny. If an idea fits one's preconceived notions, it is left unexamined.

With conspiracy theories, we thus find an odd mixture of closure and gullibility: complete acceptance of any idea that is consistent with one's ideology alongside complete rejection of any idea that is not. Conspiracy theories routinely provide insufficient evidence to survive any good scientist's criticism, yet no refutatory evidence whatsoever seems enough to convince the conspiracy theorist to give up their pet theory. This is charlatanism of the highest order—in some ways, the very opposite of science.

It is always fun to try to work out such hard and fast distinctions as we see in this matrix. I would argue, however, that there is something wrong—or at least incomplete—about it. In practice, denialists are not quite so skeptical and pseudoscientists are not quite so open. Both seem guided by a type of ideological rigidity that eschews true openness or skepticism, and instead seems to have much more in common with conspiracy theories. Although Sagan's insight is provocative, and can be used as a stalking horse, the real problem with both denialism and pseudoscience is their lack of the scientific attitude.

Denialists Are Not Really Skeptics

Denialists are perhaps the toughest type of charlatans to deal with because so many of them indulge in the fantasy that they are actually embracing the highest standards of scientific rigor, even while they repudiate scientific standards of evidence. On topics like anthropogenic climate change, whether HIV causes AIDS, or whether vaccines cause autism,[20] most denialists really don't have any other science to offer; they just don't like the science we've got.[21] They will believe what they want to believe and wait for the evidence to catch up to them. Like their brethren "Birthers" (who do not accept Barack Obama's birth certificate) or "Truthers" (who think that George W. Bush was a co-conspirator on 9/11), they will look for any excuse to show that their ill-warranted beliefs actually fit the facts better than the more obvious (and likely) rational consensus. While they may not actually care about empirical evidence in a conventional sense (in that no evidence could convince them to give up their beliefs), they nonetheless seem eager to use any existing evidence—no matter how flimsy—to back up their preferred belief.[22] But this is all based on a radical misunderstanding or misuse of the role of warrant in scientific belief. As we know, scientific belief does not require proof or certainty, but it had better be able to survive a challenge from refuting evidence and the critical scrutiny of one's peers. But that is just the problem. Denialist hypotheses seem based on intuition, not fact. If a belief is not *based* on empirical evidence, how can we convince someone to *modify* it based on empirical evidence? It is almost as if denialists are making faith-based assertions.

Unsurprisingly, most denialists do not see themselves as denialists and bristle at the name; they prefer to call themselves "skeptics" and see themselves as upholding the highest standards of science, which they feel have been compromised by those who are ready too soon to reach a scientific conclusion before all of the evidence is in. Climate change is not "settled science," they will tell you. Liberal climate scientists around the world are hyping the data and refusing to consider alternative hypotheses, because they want to create more work for themselves or get more grant money. Denialists customarily claim that the best available evidence is fraudulent or has been tainted by those who are trying to cover something up. This is what makes it so frustrating to deal with denialists. They do not see themselves as ideologues, but as doubters who will not be bamboozled by the poor scientific reasoning of others, when in fact they are the ones who are succumbing to highly improbable conspiracy theories about why the available evidence is insufficient and their own beliefs are warranted despite lack of empirical support. This is why they feel justified in their adamant refusal to change their beliefs. After all, isn't that what good skeptics are supposed to do? Actually, no.

Skepticism plays an important role in science. When one hears the word "skepticism" one might immediately think of the philosopher's claim that one cannot know anything; that knowledge requires certainty and that, where certainty is lacking, all belief should be withheld. Call this *philosophical skepticism*. When one is concerned with nonempirical beliefs—such as in Descartes's *Meditations*, where he is concerned with both sensory *and* rational belief—we could have a nice discussion over whether fallibilism is an appropriate epistemological response to the wider quest for certainty. But, as far as science is concerned, we need not take it this far, for here we are concerned with the value of doubt in obtaining warrant for empirical beliefs.

Are scientists skeptics? I believe that most are, not in the sense that they believe knowledge to be impossible, but in that they must rely on doubt as a crucible to test their own beliefs before they have even been compared to the data. Call this *scientific skepticism.*[23] The ability to critique one's own work, so that it can be fixed in advance of showing it to anyone else, is an important tool of science. As we have seen, when a scientist offers a theory to the world one thing is certain: it will not be treated gently. Scientists are not usually out to gather only the data that support their theory, because no one else will do that. As Popper stated, the best way to learn whether a theory is any good is to subject it to as much critical scrutiny as possible *to see if it fails.*

There is a deeply felt sense of skepticism in scientific work. What is distinctive about scientists, however, is that unlike philosophers, they are not limited to reason; they are able to test their theories against empirical evidence.[24] Scientists embrace skepticism both by withholding belief in a theory until it has been tested and also by trying to anticipate anything that might be wrong in their methodology. As we have seen, doubt alone is not enough when engaging in empirical inquiry; one must be open to new ideas as well. But doubt is a start. By doubting, one is ensuring that any new ideas are first run through our critical faculties.

What of scientists whose skepticism leads them to reject a widely supported theory—perhaps because of an alternative hypothesis that they think (or hope) might replace it—but with no empirical evidence to back up the idea that the current theory is false or that their own is true? In an important sense, they cease to be scientists. We cannot assess the truth or likelihood of a scientific theory based solely on whether it "seems" right or fits with our ideological preconceptions or intuitions. Wishing that something is true is not acceptable in science. Our theory must be put to the test.[25]

And this is why I believe that denialists are not entitled to call themselves skeptics in any rightful sense of the word. Philosophical skepticism is when we doubt everything—whether it comes from faith, reason, sensory evidence, or intuition—because we cannot be certain that it is true. Scientific skepticism is when we withhold belief on empirical matters because the evidence does not yet allow us to meet the customarily high standards of justification in science. By contrast, denialism is when we refuse to believe something—even in the face of what most others would take to be compelling evidence—because we do not want it to be true. Denialists may use doubt, but only selectively. Denialists know quite well what they hope to be true, and may even shop for reasons to believe it. When one is in the throes of denial, it may feel a lot like skepticism. One may wonder how others can be so gullible in believing that something like climate change is "true" before all of the data are in. But it should be a warning sign when one feels so self-righteous about a particular belief that it means more than maintaining the consistent standards of evidence that are the hallmark of science.

As Daniel Kahneman so eloquently demonstrates in his book *Thinking Fast and Slow*, the human mind is wired with all sorts of cognitive biases that can help us to rationalize our preferred beliefs.[26] Are these unconscious biases perhaps the basis for denialism even in the face of overwhelming evidence? There is good empirical support to back this up.[27] Furthermore, it cannot be overlooked that the phenomenon of "news silos" that we spoke of earlier may exacerbate the problem by giving denialists a feeling of community support for their fringe beliefs. Yet this opens the door to a kind of credulousness that is anathema to real skeptical thinking.

In fact, denialism seems to have much more in common with conspiracy theories than with skepticism. How many times have you heard a conspiracy theorist claim that we have not yet met a sufficiently high standard of evidence to believe a well-documented fact (such as that vaccines do not cause autism), then immediately exhibit complete gullibility that the most unlikely correlations are true (for instance, that the CDC paid the Institute of Medicine to suppress the data on thimerosal)? This fits completely with the denialist pattern: to have impossibly high standards of proof for the things that one does not want to believe and extremely low standards of acceptance for the things that fit one's ideology. Why does this occur? Because unlike skeptics, denialists' beliefs are not borne of caring about evidence in the first place; they do not have the scientific attitude. The double standard toward evidence is tolerated because it serves the denialists' purpose. What they care about most is protecting their beliefs. This is why one sees all of the cheating on scientific standards of evidence, even when empirical matters are under discussion.

The matrix that I concocted from Sagan's work therefore seems wrong in three important ways about denialism.[28] First, it seems wrong to classify denialists as skeptics. They may use evidence selectively and pounce on the tiniest holes in someone else's theory, but this is not because they are being rigorous; the criteria being used here are ideological, not evidential. To be selective in a biased way is not the same thing as being skeptical. In fact, considering most of the beliefs that denialists prefer to scientific ones, one must conclude that they are really quite gullible.[29] Second, it also seems wrong to say that denialists are always closed to new ideas. As we will see in the example of climate change, denialists are plenty open to new ideas—and even empirical evidence—when it supports their preexisting beliefs. Finally, there may be an error in Sagan's contrast between skepticism and openness. Sagan argues that these two notions must be balanced in scientific reasoning, which implies that they are somehow in conflict. But are they? In Massimo Pigliucci's *Nonsense on Stilts*, he observes that

> to be skeptical means to harbor reasonable reservations about certain claims. … It means to want more evidence before making up one's mind. Most importantly, it means to keep an attitude of openness, to calibrate one's beliefs to the available evidence.[30]

I believe that this is an accurate account of the nature of scientific skepticism. How can one be open-minded enough to suspend one's belief, yet not be open to new ideas? Skepticism is not about closure; it is about forcing oneself to remain open to the possibility that what seems true may not be. Science is relentlessly critical, but this is because no matter how good one's evidence, a better theory may await over the next horizon.

Denialism in Action: Climate Change

Perhaps the best example of scientific denialism in recent years is climate change. The theory that our planet is getting progressively hotter because of the release of greenhouse gases caused by human consumption of fossil fuels is massively supported by the scientific evidence.[31] There remains, however, great public confusion and resistance to this, as a result of the various monied, political, and media interests that have whipped it into a partisan issue. The sordid story of how those with fossil fuel interests were able to capitalize on the foibles of human reasoning by "manufacturing doubt" where there was none—substituting public relations for scientific rigor—is a chilling tale of the vulnerability of science. The single best book on this is Naomi Oreskes and Erik Conway's *Merchants of Doubt*.[32] In my own *Respecting Truth*, I engage in an extended discussion of the epistemological fallout that resulted from public confusion not only over whether global warming is true, but also over whether the vast majority of *scientists* believe that it is true (which they do).[33]

Some of the most intellectually irresponsible rhetoric has come from politicians who have tried to impugn the reputation of climate scientists by calling climate change a hoax.[34] One sometimes wonders whether they really believe this, or are just "paying the crazy tax" of trying to get elected in an environment in which a frightening percentage of the public believes it; but either way this is a shameful self-stoking cycle. The more politicians lie, the more these lies are reflected in public opinion.

One of the worst perpetrators is US Senator Ted Cruz. At an August 2015 campaign event, sponsored by the Koch Brothers, Cruz said this:

> If you look at the satellite data in the last 18 years there has been zero recorded warming. Now the global warming alarmists, that's a problem for their theories. Their computer models show massive warming that satellites says ain't happening. We've discovered that NOAA, the federal government agencies are cooking the books.[35]

What's wrong with this statement? Well, for one thing it isn't true. This idea of a "global warming hiatus" has been around for years but was recently disproven by Thomas Karl, director of the NOAA's National Centers for Environmental Information, in an article in *Science* in June 2015.[36] To give Cruz the benefit of the doubt, perhaps he had not known of Karl's article at the time of his speech. Yet Cruz did not apologize and retract his statement later, after Karl's article was well publicized. Indeed, in December 2015, Cruz sat for a remarkable interview with NPR, which is so enlightening of the denialist mindset it is worth quoting at length:

> **Steve Inskeep, Host:** What do you think about what is seen as a broad scientific consensus that there is man-caused climate change?
>
> **Ted Cruz:** Well, I believe that public policy should follow the science and follow the data. I am the son of two mathematicians and computer programmers and scientists. In the debate over global warming, far too often politicians in Washington—and for that matter, a number of scientists receiving large government grants—disregard the science and data and instead push political ideology. You and I are both old enough to remember 30, 40 years ago, when, at the time, we were being told by liberal politicians and some scientists that the problem was global cooling.
>
> **Inskeep:** There was a moment when some people said that.
>
> **Cruz:** That we were facing the threat of an incoming ice age. And their solution to this problem is that we needed massive government control of the economy,

the energy sector and every aspect of our lives. But then, as you noted, the data didn't back that up. So then, many of those same liberal politicians and a number of those same scientists switched their theory to global warming.

Inskeep: This is a conspiracy, then, in your view.

Cruz: No, this is liberal politicians who want government power over the economy, the energy sector and every aspect of our lives.

Inskeep: And almost all the countries in the world have joined in to this approach?

Cruz: So let me ask you a question, Steve. Is there global warming, yes or no?

Inskeep: According to the scientists, absolutely.

Cruz: I'm asking you.

Inskeep: Sure.

Cruz: OK, you are incorrect, actually. The scientific evidence doesn't support global warming. For the last 18 years, the satellite data—we have satellites that monitor the atmosphere. The satellites that actually measure the temperature showed no significant warming whatsoever.

Inskeep: I'll just note that NASA analyzes that same data differently. But we can go on.

Cruz: But no, they don't. You can go and look at the data. And by the way, this hearing—we have a number of scientists who are testifying about the data. But here's the key point. Climate change is the perfect pseudoscientific theory for a big government politician who wants more power. Why? Because it is a theory that can never be disproven.

Inskeep: Do you question the science on other widely accepted issues—for example, evolution?....

Cruz: Any good scientist questions all science. If you show me a scientist that stops questioning science, I'll show you someone who isn't a scientist. And I'll tell you, Steve. And I'll tell you why this has shifted. Look in the world of global warming. What is the language they use? They call anyone who questions the science—who even points to the satellite data—they call you a, quote, "denier." Denier is not the language of science. Denier is the language of religion. It is heretic. You are a blasphemer. It's treated as a theology. But it's about power and money. At the end of the day, it's not complicated. This is liberal politicians who want government power.

Inskeep: You know that your critics would say that it's about power and money on your side. Let's not go there for the moment. But I want to ask about this. I want to ask about facts.

Cruz: But hold on a second. Whose power—but let's stop. I mean, if you are going to ...

Inskeep: Energy industry, oil industry, Texas ...

Cruz: If you're going to toss an ad hominem.[37]

There are so many possible things to find fault with here that it is almost a textbook case of poor reasoning: the double standard of evidence, the subtle change of subject when he was pinned down on conspiracy theories, deliberate misunderstanding of what the "openness" of science amounts to, and the schoolyard rhetorical trick of "I know you are, but what am I?" Let us focus, however, on the one empirical claim that was repeated about the alleged eighteen-year pause in global warming. As it turns out, the government's statistics on climate change suit Cruz just fine when they show something he likes. In this case, it was an (erroneous) IPCC assessment report from 2013 (which has since been corrected).[38] This happens sometimes in science; errors are made and they need to be corrected, but not because there is a conspiracy.[39] So Cruz is using an outdated, incorrect, discredited graph. But there's another problem too. Eighteen years is a weird number. Notice that Cruz didn't say "in the last twenty years" or even "in the last seventeen" or "in the last nineteen." Why would he be so specific? Here one must think back to what was happening exactly eighteen years prior to 2015: El Nino.

Here we encounter the denialist's penchant for cherry picking evidence. Despite the fact that fourteen of the last fifteen years had yielded the hottest temperatures of the century, 1998 was a (high) outlier even among those. It showed an astonishingly high pickup in global temperatures for that year only. If you think about the graph that might accompany these data, you can imagine that choosing such a high-temperature year as your base point would make 2015 look relatively cooler. When examined out of context, the eighteen-year gap between 1998 and 2015 made it look like the global temperature had been fairly flat. But it wasn't. As we now know from Karl's study, some of those temperature results were not only wrong but, as any scientist can tell you, you've also got to look at the *whole graph*—including the years in between—which show that 1998 was the high outlier and that there has been a steady trend in global warming over the last several decades.[40] Even if one uses the old uncorrected graph, Cruz's reasoning is flawed.

Cherry picking data is a cardinal offense against the scientific attitude, yet it is a common tactic of denialists. Few scientists would make this error. In science, one's ideas must be subjected to rigorous tests against reality based on previously accepted standards; one can't just pick and choose as one goes. But to an ideologue like Ted Cruz (and, as it turns out, to many who are not trained to avoid this type of error in reasoning), it may feel perfectly natural to do this. The reason is explained by cognitive psychologists and behavioral economists in their work on confirmation bias and motivated reasoning. As we've seen, confirmation bias is when we seek out reasons to think that we are right. Motivated reasoning is when we allow our emotions to influence the interpretation of those reasons relentlessly in favor of what we already think. And both of these are completely natural, built-in cognitive biases that are shared by all humans, even when they have been trained to guard against them. Scientists, given their education in statistics—and the fact that science is a public enterprise in which proof procedures are transparently vetted by a community of scholars who are looking for flaws in one's reasoning—are much less prone to make these kinds of errors. Those who have been trained in logic, like philosophers and others who take skepticism seriously, can also be expected to recognize these biases and guard against their stealthy erosion of good reasoning. But most denialists? Why in the world should they care?

Of course, few denialists would agree with this assertion, primarily because they would deny that they are denialists. It sounds so much more rigorous and fair-minded to maintain their "skepticism," which probably accounts for the recent hijacking of this word.[41] In fact, some denialists (witness Cruz) go so far as to claim that *they* are the ones who are really being scientific in the climate change debate. The claim is a familiar one. The science is "not yet settled." There is "so much more" that we need to know. And isn't climate change "just a theory"? But the problem is that none of this is based on a good faith commitment to any sort of actual skepticism. It is instead based on a grave misunderstanding of how science actually works coupled with a highly motivated capitulation to cognitive bias. Yes, it is true that the science of climate change is not *completely* settled. But, as we have seen, that is because *no* science is ever *completely* settled. Given the way that science works—which we explored in chapter 2—it is always going to be the case that there are more experiments we can do or tests we can run. But it is a myth to think that one needs complete confirmation or proof before belief is warranted. (And indeed if the denialist rejects this, then why the double standard for their own beliefs?)

As for the claim that climate change is "just a theory"—so any alternative theory "could be true"—one is reluctant to give this much credence. As previously noted, gravity is just a theory. So is the germ theory of disease. As we have seen, some scientific theories are exceptionally robust. But the standard of warrant for scientific belief is *not* that anything goes until it has been disproven. While it is correct to say that, until it is refuted by the evidence, any theory *could* technically speaking be true, this does not mean that the theory is justified. Nor is it worth scientists' time to run every fringe hypothesis to ground. Cranks should not expect to be able to kick the door in and insist that, because science is supposed to be open-ended, their theory must be tested. Science is rightly selective, and the criterion must be warrant, based on the evidence.

Scientific explanation is not made up of correct guesses or having just a few data points. Consider the Flat Earth hypothesis. If it is true, where is the evidence? Since there is none, one is scientifically justified in disbelieving it. Flat Earthers[42] are customarily reluctant to say what is wrong with the evidence in favor of heliocentrism, other than that until it has been "proven," their own theory *could* be true. But that is not correct reasoning. Even if someone guessed correctly at something that turned out to be true, this is not science. Science is about having a theory that backs up one's guesses; something that has been tested against—and fits with—the evidence.

Another denialist misunderstanding occurs over how scientists reach consensus. Again, one does not need 100 percent agreement before the field moves on. Those who are asking for complete agreement among the world's scientists before something is done about climate change are just running out the clock. According to the latest statistics, over 96.2 percent of the world's climate scientists believe that climate change is occurring and that humans are responsible for it.[43] For comparison, note that a similar survey found that only 97 percent of scientists believe in evolution by natural selection, which is a bedrock principle of biology.[44] More than one hundred and fifty years after Darwin, we still do not have 100 percent agreement even on evolution! But we don't need it, because that is not how scientific consensus works. Scientific claims are subjected to group scrutiny and criticism and, despite inevitable dissent, the field makes a choice.[45] Some may complain that this still leaves room for doubt and that the "skeptics" could be right (as they sometimes are), but I wouldn't hold out much hope for denialism about climate change. For one thing, legitimate dissenters must be able to present some empirical evidence as reason for their dissent. Without that, one questions whether they are even scientists. Denialists might complain that they *do* have evidence and that they *are* members of a community, yet this is simply tiresome. As we've seen, seeking out a community for mutual agreement is not the same thing as seeking out critical scrutiny. No matter how many people you get to agree with you, majority opinion does not trump evidence in a factual matter.[46]

Couldn't the scientists nonetheless be wrong? Yes, of course. The history of science has shown that *any* scientific theory (even Newton's theory of gravity) could be wrong. But this does not mean that one is a good skeptic merely for disbelieving the well-corroborated conclusions of science. To reject the cascade of scientific evidence that shows that the global temperature is warming, and that humans are almost certainly the cause of it, is not good reasoning *even if some long-shot hypothesis comes along in fifty years to show us why we were wrong.* Skepticism is in complete conformity with the scientific attitude's commitment to form one's beliefs based on fit with the evidence, and then change them as better evidence comes in, but this in no way guarantees truth. Science instead offers justification based on the evidence. Yet this is a mighty thing. With denialism, one decides in advance—on the basis of ideology—what one wants to be true, then selectively filters information based on whether it supports the hypothesis or not. But this does not build warrant. Science may sometimes miss the mark, but its successful track record suggests that there is no superior competitor in discovering the facts about the empirical world.

Indeed, this is why someone like Galileo was not a denialist. Who claimed that he was? In an interview with the *Texas Tribune* in March, 2015, Ted Cruz said:

> Today the global warming alarmists are the equivalent of the flat-Earthers. It used to be ... accepted scientific wisdom the Earth is flat, and this heretic named Galileo was branded a denier.[47]

We can make pretty short work of this claim. Galileo believed in the heliocentric model not because of ideology but because *of evidence.* His telescopic observations of the phases of Venus, craters on the Moon, and the satellites of Jupiter provided a mountain of reasons for believing that the Ptolemaic model of an Earth-centered universe was wrong, and it should have convinced anyone who doubted him. The Catholic Church was the one with ideological beliefs that prevented them from accepting the reality of Galileo's evidence because of what it would have meant for their celestial model. Galileo most certainly was *not* a denier. To deny the truth of a false theory *when you have the evidence to show that it is false* is not denialism; it is science.

What might happen when the lone dissenter *does* have the evidence? When that person questions the established consensus of science and shows that it is wrong? This challenges the idea that science is a privileged process resulting from group scrutiny of individual work. Can the scientific attitude survive intact if someone goes against the scientific community—which is supposed to be the final arbiter of justificatory warrant—and wins?

What Happens When the "Crank" Is Right?

J Harlen Bretz was a maverick geologist in the early twentieth century, who spent a long career at the University of Chicago, but did his fieldwork in a desolate region of Washington state that he termed the "channeled scablands." This area is remarkable for its severe "Mars-scape" surface, consisting of washed-out channels surrounded by high bluffs (where gravel and "erratic rocks" are found thousands of feet above sea level), U-shaped canyons with dry falls, and enormous plunge pools with only small waterfalls draining into them. It is, in short, a geography that demands an explanation.

This area had not been well studied before Bretz, but geologists of the time did have their hypotheses. Most agreed that this landscape had been carved by hydrologic forces, but—in keeping with the "uniformitarian" theory of the time—they thought that this must have been the result of relatively small amounts of water acting over long periods of time, such as those that had created the Grand Canyon. Uniformitarianism, which had been the dominant paradigm in geology at least since Charles Lyell's influential textbook (which inspired Darwin's theory of evolution by natural selection) is the idea that the geological record can be explained by known natural forces acting over millions of years.[48] This was proposed in contrast to the catastrophism of Lyell's predecessors, who felt that short-term cataclysmic events—perhaps caused by God—had created the geological, fossil, and biological record. Natural forces versus miracles. Erosion versus a catastrophe. It was not hard to figure out which side most scientists would be on.

Although Bretz was himself a uniformitarian (and an atheist), as he stood before the scarred, barren landscape for the first time in 1921, he began to realize that the prior theories must be wrong. Like a detective solving a mystery, Bretz found too many clues that this could not have been the result of steady erosion. What was the alternative? A massive flood. A flood so large that it would have been thirteen miles wide at some points and involved a volume of water so strong it could force the Snake River to flow backward, make U-shaped channels rather than V-shaped ones as far south as the Columbia Gorge, put the "turtle on the fencepost" of gravel found on top of 2,500-foot-high bluffs, and create the enormous plunge pools that were so disproportionate to the tiny falls that fed into them. Where could such an enormous amount of water have come from? Bretz did not know and, for the moment, he framed no hypotheses. It was a puzzle and he vowed merely to follow the evidence wherever it landed him. By the time he came back for the second summer to continue his fieldwork, he was convinced:

> Bretz now believed that these geologic features could only have been created by a flood of unimaginable proportions, possibly the largest flood in the history of the world. And this was no claim made as wild speculation. Fact after fact, feature after feature in the landscape had proven to Bretz that his theory provided the only plausible answer for the formation of the channeled scablands.[49]

The story of Bretz's work is a fascinating and under-reported topic in the history and philosophy of science, which would repay much greater scholarly attention. John Soennichsen's *Bretz's Flood* is one of the few available resources, but fortunately it is wonderful, both for telling the story of Bretz's struggles but also providing the intellectual context of the time with discussion of geological positivism, uniformitarianism versus catastrophism, the struggle to make geology more scientific, and how sociological factors can influence scientific explanations. Here I will focus on the narrow issue of what implications Bretz's story might have for the scientific attitude, when group consensus is against the individual who is right and has the evidence. Does this challenge the idea that the scientific attitude is instantiated at the level of group practice and that science works primarily because the community of scientists corrects the errors of individuals? As I hope to show below, I think that the Bretz episode not only survives this challenge but is in fact a stunning demonstration of the power of the scientific attitude.

It is obvious that Bretz's work is an endorsement of the scientific attitude at the individual level. He gathered enormous amounts of evidence to back up his hypothesis and critiqued and shaped his theory along the way. Because his theory was in some ways a throwback, he understood that it would be savagely attacked.[50] Bretz did not posit any supernatural forces to explain the geological record, but neither could he explain it with currently existing forces acting over a long period of time that uniformitarianism demanded. Instead, he had to propose an enormous catastrophic event that had occurred over a short period, with an unknown source of water. It would sound like something out of the Bible; the push back would be tremendous.

> Geology, like other sciences, is a brotherhood of sorts, with much camaraderie and the sharing of data, ideas, and facilities. It is seen as a field in which the work of one individual can inspire another, and through the cooperation of all, fledgling theories can expand and flower into fully matured scientific views shared by the discipline as a whole. As with other brotherhoods, however, squabbles can erupt and family members who don't adhere to the basic rules can be verbally disciplined or—perhaps worse—ignored altogether.[51]

What Bretz was proposing sounded like heresy. To Bretz's credit, he didn't seem to care; he had the evidence and knew he was right. The others would just have to come around. In this, Bretz reminds one of a modern-day Galileo.[52] Those who fought him had not done the fieldwork or seen the landscapes.[53] Bretz was stubborn enough to believe in the implications of what he had seen firsthand, and if it did not square with current theories, then those theories must be wrong. This is good testament to Bretz's adherence to the scientific attitude, but there was trouble ahead.

One problem for Bretz was that he still had no idea what could have caused such a megaflood. The amount of water required would be tremendous. He would be telling a causal story with no cause in sight, and he knew this would present a formidable barrier to acceptance of his theory. But the geologic record simply demanded that amount of water. Here Bretz also reminds one of Darwin, who gathered evidence that supported his theory of evolution by natural selection long before he had any idea of the mechanism that might be behind it. One is also reminded of Newton, who "framed no hypothesis" about what gravity was, as he worked out the equations that described its behavior.

It's important here to pause and consider the implications for the scientific attitude, because it reveals that understanding a cause is not necessary for scientific explanation. Causal explanations are an important part of complete scientific theories, and positing a miracle is not an acceptable substitute. But most important is having evidence that a hypothesis is warranted; cause can be inferred later. This is not to trivialize causal explanation. It's just that finding a cause is often the last part of scientific explanation, coming into place after all of the other pieces of evidence have been gathered and fit together.

In Bretz's case, it would have been preferable if he had known a cause, but initially he simply could not specify one, so he stuck to his evidence from the geologic record. Note that Bretz was enormously self-critical and made numerous corrections and modifications to his theory as he went along.[54] This was nothing, however, compared to the criticism he was to receive at the hands of his fellow scientists.

In 1927, Bretz stood on the steps of the Cosmos Club in Washington, DC, to present his evidence before a gathering of the nation's most distinguished geologists. Foremost among these were six members of the US Geological Survey (USGS), who were a sort of governing board for the profession. Bretz gave a lengthy, definitive, presentation based on his six years of fieldwork in the area. Then, when it came time for the respondents to speak, "all hell broke loose."

> One by one, each of the men sitting at the presentation table rose to confront him with objections, criticisms, and—for the first time—their own interpretations of the scablands. It quickly became clear that this had been a planned attack; a strategic event that allowed Bretz to offer his views, then be subjected to the collective bile of virtually every prominent geologist of his time. … It seems clear that the official stance of that influential body was one of intolerance for any theory that strayed from the uniformitarian line.[55]

This, of course, is not how science is supposed to work. Looking back, one suspects that the learned scientists were objecting to Bretz's theory because of motivated reasoning rooted in a near-ideological commitment to uniformitarianism. It is one thing to attack a theory based on lack of fit with the evidence; it is another to say that it cannot be tolerated because it would erode the field's hard-won progress in distancing itself from religious views that would keep it from being seen as scientific.

In this book, I have made the case that the scientific attitude is instantiated not just in the values of individual scientists, but in the community of scholars. So what can be said about a case where the "heretic" is behaving more scientifically than his critics? Is it possible to defend the importance of the scientific attitude even when the consensus of one's profession is wrong? This is a delicate question, for although it is common for individual scientists sometimes to outpace the group—indeed, this is often how breakthrough ideas come to disseminate and lead to a changed understanding—what is rare is for someone to openly contradict the scientific consensus for decades and later be vindicated. Great historical examples of this include Galileo, Semmelweis, Alfred Wegener, and other martyrs for science. But when such episodes occur, what is the guidepost for keeping science on track? It has to be the evidence. It's not that Bretz was making some wild claim out of the blue: he had the evidence to back it up. So even while it is understandable that there will sometimes be community resistance to those are perceived to have departed from orthodoxy in their scientific thinking, over time this conflict must be resolved in the only way possible in science: through empirical evidence.

And this was in fact precisely what occurred with Bretz's theory of the channeled scablands. After the disastrous Cosmos Club talk, one of Bretz's previous rivals helped him to realize that the only water source big enough for such a massive flood must have been the spontaneous draining of a glacial lake. This turned out to be the right answer. It is now thought that the failure of a giant ice dam at Lake Missoula released over 500 cubic miles of water that reached 100 miles south of Portland, Oregon.[56] Still the detractors held on.

> The geologic community went about their business and tried to ignore the fact that this upstart geologist was spouting nonsense about massive floods he claimed had altered the topography of a vast western landscape in a geologic blink of an eye.[57]

Bretz went into a deep depression. As time passed, some of his critics eventually died, while others came around, but a new generation of geologists also grew up who were more sympathetic to Bretz's theory.[58] After decades, Bretz was eventually vindicated. As one of his critics said years later when he finally visited the scablands himself, "How could anyone have been so wrong?"[59] How sweet it must have been in 1965 when a group of geologists made an official field trip to the scablands and sent Bretz this telegram: "We are all catastrophists now."[60]

In a strange coda to Bretz's story, his legacy has now been claimed by thousands of creationists, who regard him as a sort of folk hero for almost single-handedly proving their case for a biblical flood. Of course, Bretz did no such thing, yet there are creationist websites on which he is celebrated as a David who went up against the organized science Goliath and won.[61] What to say about this? And does Bretz's example give support to the idea that someone like Ted Cruz just *may* be the next Galileo? This seems preposterous; remember that the lodestar in Bretz's story is one of sticking to the evidence and eschewing ideology. If a science denier claims that climate change is a hoax, where is the evidence? Without this, it is just speculation or worse. This is not to say that simple skepticism or even stubbornness marks a departure from science. The standards for accepting a new theory should be high. But when a scientist abandons evidence for ideology, this is no longer science.

What might happen if a purportedly crackpot theory *did* have the evidence? Then it must be tested and, if it holds up, the scientific consensus must change. Just as the scientific attitude should ideally guide individual behavior, so too it must guide group behavior. Science is supposed to be self-correcting at both the individual and the group level. Imagine a scientist with an outlier theory that did not square with the evidence and was rejected by the rest of the profession. If this scientist clings to the theory, in some sense he or she will have left the profession. Can the same thing happen to an entire discipline? Although it is more commonly the case that the group corrects the individual—as in the case of cold fusion—it does sometimes occur that the individual corrects the group.

Just as geology was temporarily taken off course by refusal to accept Alfred Wegener's theory of continental drift, so it was later to suffer the same fate with Bretz's theory of the channeled scablands. It is painful to think that geology was "not scientific" for a time, but that is as it must be when scientists refuse to change a theory in the face of compelling evidence. Just as the Catholic Church refused to acknowledge the truth of Galileo's new theory of the heavens—choosing instead to stick with its familiar but false ideology—so for a time did geology choose to embrace strict uniformitarianism over empirical evidence.[62] What makes science distinctive, however, is that even when this occurs, there is a way to get back on track.[63] Indeed, note that—as a science—geology did eventually recognize the force of Bretz's data, and return to the scientific attitude. (The Catholic Church, however, did not and was instead humiliated into an apology to Galileo 350 years after it had already lost the debate over heliocentrism.)

We must here face squarely the implications of what this example says about the question of whether the scientific attitude is a defining feature of science. The idea behind the scientific attitude cannot be that the group is always right. Galileo, Semmelweis, Wegener, and Bretz provide the counterexamples. The individual is sometimes far ahead of his or her contemporaries. True, as we saw earlier in Sunstein's experimental studies, it is often easier for groups to find the truth than for individuals. But this does not mean that this always happens. Sometimes the individual has the better theory. And that is perfectly all right in science. What is important for the preservation of the scientific attitude is that any disputes are settled by the evidence. Sometimes it is not just individual theories that need correction, but an entire discipline's consensus.

Returning to Bretz, it is worth considering for a moment *why* uniformitarianism had such a hold on geology. In this case, I think we see one of the rare instances where an entire scientific field was subject to ideological influence. One reason that uniformitarianism was so favored by geologists was that it was seen as a bulwark against the creationists. It was a way of vindicating the idea that the natural world could be explained through the slow workings of natural processes rather than some sort of catastrophe that might be expected from divine intervention. Still, it is perfectly consistent to think that natural events can occur suddenly, over short periods of time. But this does not necessarily mean that God exists.[64] It is also important to realize that Bretz did not jump to the conclusion of catastrophism rashly. His own guiding philosophy was uniformitarianism *until he was pushed elsewhere by the evidence*. In his papers and talks on the subject, it is clear that Bretz did not take the implications of his theory lightly. He anticipated criticism and tried to address it, but still believed in his own theory because there was no other explanation for what he saw. Contrast that with some of the geologists who had not even seen the evidence of the scablands but were nonetheless trying to shut him down. In this, they were being ideologues. As scientists, why were they more committed to the theory of uniformitarianism than to following the evidence? This demands an explanation too.

My hypothesis is that ideology can have a corrupting influence both on those who are committed to it and on those who are fighting it. As scientists, the members of the USGS should not have cared whether Bretz's evidence was consistent with one overarching theory or another, yet they did. Why? Because they were locked in their own battle with fundamentalist Christians and did not want Bretz's theory to give aid and comfort to their enemies. This is an example of how ideology can infect the scientific process, whether we are on the "right" side of it or not. That is, even if we are merely changing the way that we do science to fight against those who are *unscientific*, we can damage science. It is not a question of whether Bretz's theory came from a lone individual or a group, whether it advocated gradual change or sudden intervention. What matters is that it was consistent with the data. But when we try to go around this—when we seek either to confirm or disconfirm a theory based on our extra-empirical commitments—problems can arise. Most often this occurs when religious or political ideologues crab the process of learning from evidence because of their personal beliefs about divine intervention, human freedom, equality, nature, nurture, or some other speculative commitments. But this can also occur when an individual or group is fighting *against* such ideologies as well. The temptation to push things a little toward the way that we think (or hope) truth lies can be great. But this can result in unexpected reversals (or even frauds), which can then backfire and do a disservice to the public's trust in science.

Consider here the "climate-gate" email controversy from a few years back, where a handful of scientists spoke of suppressing evidence in response to Freedom of Information Act requests that they knew would be used by denialists to cherry pick data and undercut the truth about global warming. Although they were joking, and surely must have felt that they were on the "right side" of science, the fallout for climate science was terrible. Even after multiple official inquiries—which showed that the scientists had actually done nothing wrong and that their work was never compromised—it fed straight into the conspiracy theories of those who had argued that global warming was a hoax perpetrated by liberal scientists. When we compromise our standards, even if we feel that we are working "on the side of the angels," science can suffer.[65]

It is deeply frustrating to have science attacked by ideologues, who care nothing for what is precious about it and seek data only to support their favored hypotheses. But the price of scientific freedom is eternal openness. This does not mean that we have to tolerate crackpot theories. If there is no evidence, and no warrant behind them, there is no reason to put the scarce resources of science into checking them. But what to do when the evidence *does* show something strange? Here Sagan seems right. We must give it a hearing. And this is exactly what we will do at the end of this chapter when we consider the nearly three decades of work done on ESP at the Princeton Engineering Anomalies Research (PEAR) center. But first we must attend to the topic of pseudoscience.

Pseudoscientists Are Not Really Open to New Ideas

The problem with pseudoscientists is not merely that they are not doing science, but that they claim that they are. Some probably know that they are only pretending. Others perhaps believe that their work is being unfairly scorned. But the bottom line is that when one is making explanatory claims about empirical matters, fit with the evidence should be the primary consideration.[66] Why then is it so difficult to get most pseudoscientists to admit not just that their theories are not true, but that they are not even scientific? The answer is that, similar to denialists, their belief in their theories seems deeply rooted in wishful thinking.

Clearly, wishful thinking is not in keeping with the scientific attitude. One should not decide in advance on the basis of ideology what one wants to be right, then chase the evidence to support it. In science, one is supposed to be guided by the evidence and one's beliefs are supposed to be shaped by it. As we know, scientific hypotheses can come from anywhere. Intuition, wishful thinking, hope, stubbornness, and wild guesses have all resulted in respected scientific theories. But here is the key: they must be supported by the evidence, as judged by the community of other scientists.

Here once again consider Sagan's matrix. Are pseudoscientific hypotheses open to new ideas? Not particularly. While it is probably fair to say that many astrologers, dowsers, crystal healers, intelligent design theorists, and the like are extremely gullible (as Sagan notes), they are also customarily closed-minded to an extreme degree in accepting the import of any evidence that disproves their theories. They will not submit to falsification. Controlled experiments are rare. Cherry picking data is common. Like denialists, most pseudoscientists seem to want to avoid disconfirming evidence, even as they complain that other scientists will not consider their own.

One expects that some are profiting from this cat-and-mouse game and know what they are doing. While some are misleading, others are misled. Astrology is a billion-dollar industry worldwide.[67] According to NBC News, Americans spend $3 billion a year on homeopathy.[68] Other advocates of pseudoscience are surely straight-up ideologues who are in it not for the money but because they think they are right. And of course there is always willful ignorance and those who are duped. All are a danger to good science. Whether someone actually believes their untruths or merely pretends to, it is hostile to science to refuse to form empirical beliefs based on commitment to the standards of evidence. In pseudoscience as in denialism, they eschew (or at least cheat on) the scientific attitude. Intuition is prized over fact. "Skepticism" is used at one's convenience. Gullibility is rampant. A double standard is applied to evidence. Dark conspiracies are spun about the work of those who oppose them. Both pseudoscience and denialism surely also include those who are benefiting from public confusion, while others naively do their bidding.[69]

The crucial question for pseudoscientists is this: If your theories are true, where is the evidence? You may claim that you are being persecuted or ignored by mainstream science, but if you actually have good answers, why would they do that? As we just saw with the example of Harlen Bretz, if you have the evidence, the rest of the field will *eventually* beat a path to your door. But the onus is still on you. If even an eminent scientist like Bretz faced fierce, sometimes unreasoned opposition to his theory, why should pseudoscientists expect to have it easier? It is perhaps not great testimony for open-mindedness in the practice of science that Bretz had to fight so hard—even when he had the evidence—but that is the plight of the maverick. Scientists are stingy with warrant. So why do pseudoscientists expect to be taken seriously when they can offer no—or only equivocal—evidence? Because they "might" be right? But we have already seen that this matters little when the question of justification is on the line.

Where are the falsifiable predictions of astrology? Where are the double-blind controlled experiments by faith healers? Why have those who claim that time travel is possible never gone back and made a killing in the stock market?[70] If those who hold "alternative" beliefs want their views to be taken seriously, they must expect them to hold up under intense scrutiny and criticism. As we saw earlier, this sometimes happens, and the results are routinely disappointing.[71] Instead, pseudoscientists usually prefer to release their own selective evidence. But this is only a masquerade of science.

Pseudoscience in Action: Creationism and Intelligent Design Theory

The long sordid history of opposition to evolutionary theory has been well told elsewhere.[72] Starting with the Scopes Monkey Trial in Tennessee in 1925, those who sought to fight against Darwin's theory of evolution by natural selection first chose to try to keep it out of the public school biology classroom. This strategy was fairly successful until its constitutionality was challenged in 1967.[73] As we saw in chapter 2, a more modern creationist agenda then shifted from one of trying to keep evolution out of the classroom to one of lobbying for the inclusion of creationism alongside it. This began in Arkansas in 1981 with Act 590, which required teachers to give "balanced treatment" by teaching creation science alongside evolution science in biology classrooms. When this was successfully challenged on constitutional grounds in *McLean v. Arkansas*, Judge William Overton ruled that the claim that Darwinian biology was itself a "secular religion" was ludicrous and that "creation science" was not scientific in the least in that "a scientific theory must be tentative and always subject to revision or abandonment in light of facts that are inconsistent with, or falsify, the theory."[74] Thus was creation science revealed to be nothing more than pseudoscience.

Years later, the creationists regrouped under the banner of intelligent design (ID) theory, which purported to be a scientific alternative to evolution. This was the product of a "think tank" called the Discovery Institute, which was founded in Seattle, Washington, in 1990, with the agenda of promoting ID theory and attacking evolution. After a multiyear campaign of funding and promoting ideologically driven criticisms of evolution, and flooding the media with misinformation in a public relations blitz intended to raise doubts about evolution, the next court battle was fought in Pennsylvania in 2004, in a case called *Kitzmiller v. Dover Area School District*. Again, this history is recounted elsewhere,[75] but the main point is that the strategy was no longer one to get creationism or creation science taught in science classrooms, but instead to make the case for the *completely separate* scientific theory of intelligent design, which the paleobiologist Leonard Krishtalka among others has called "creationism in a cheap tuxedo."[76] This effort too went down in stunning defeat. In a judgment similar to the earlier Arkansas ruling, Judge John E. Jones found that intelligent design theory was not science and that its attacks on evolution had already been refuted by the scientific community. It had, moreover, none of its own peer-reviewed studies or evidence to offer in support of its claims. In a bold rebuke, Jones went on to scold school officials for such "breathtaking inanity" and wasting taxpayer money. He then ordered them to pay $1 million to the plaintiffs in damages.

After this, the creationists' strategy changed. With court options now deemed too dangerous, the opponents of evolution chose to try to influence the law itself. In 2008, the Discovery Institute drafted a piece of model legislation that sought to protect the "academic freedom" of teachers who felt intimidated or threatened in teaching the "full range of scientific views regarding biological and chemical evolution."[77] The language went on to identify the "confusion" created by the Dover ruling and stated that of course nothing in the act should be construed as "promoting any religious doctrine." This was nothing but a fig leaf for renewed attempts to try to get creationism into the nation's science classrooms.

After an initial defeat in Florida in 2008—where Democrats seized on ambiguities in the House language to argue that the academic freedom of teachers to cover birth control, abortion, and sex education should also be protected—the first such academic freedom bill was passed in Louisiana that same year. Though not completely modeled on the Discovery Institute's language, it was seen as a win for antiscience forces. Here legislators were careful to strip out all mention of evolution (and global warming) as examples of "controversial" theories from their original bill and renamed it the Louisiana Science Education Act. It was signed into law by Governor Bobby Jindahl and remains one of only two state academic freedom bills in the country.

Similar efforts then died in legislation in Missouri, Alabama, Michigan, South Carolina, New Mexico, Oklahoma, Iowa, Texas, and Kentucky, before another academic freedom bill was passed in Tennessee in 2012. This one purported to protect "teachers who explore the 'scientific strengths and scientific weaknesses' of evolution and climate change."[78] Soon after, in early 2013, four other states immediately followed suit: Colorado, Missouri, Montana, and Oklahoma. Indeed, Oklahoma has become the poster child for such legislation, as it had been reintroduced for every session in the state senate for the last five consecutive years. The language in these bills is similar.[79] In the latest 2016 Oklahoma Senate bill legislators sought to

> create an environment within public school districts that encourages students to explore scientific questions, learn about scientific evidence, develop critical thinking skills and respond appropriately and respectfully to differences of opinion about controversial issues.[80]

There is only one problem: disputes in science are and should be resolved on the basis of evidence, not opinion. The Oklahoma House bill in 2016 states that:

> The Legislature further finds that the teaching of some scientific concepts including but not limited to premises in the areas of biology, chemistry, meteorology, bioethics and physics can cause controversy, and that some teachers may be unsure of the expectations concerning how they should present information in some subjects such as, but not limited to, biological evolution, the chemical origins of life, global warming and human cloning.[81]

I am happy to report that these bills, along with similar ones in Mississippi and South Dakota, all failed in 2016. In recent years, similar bills have also failed in Arizona, Indiana, Texas, and Virginia. For those interested in keeping track of the fate of current and future antiscience legislation, there is a complete chronology at the website for the National Center for Science Education.[82]

It is sad commentary on public understanding of science that things have gone this far. As Thomas Henry Huxley ("Darwin's bulldog") once put it, "Life is too short to occupy oneself with the slaying of the slain more than once." But this is precisely the wrong attitude to have when one is fighting pseudoscience, which is perennial. As we have seen, the tactics shift and new strategies are employed, but the fight must go on.

Indeed, I witnessed this firsthand in my own skirmish with the Discovery Institute. In a 2015 article entitled "The Attack on Truth" that appeared in the *Chronicle of Higher Education*, I wrote that the Discovery Institute was a "Seattle organization advocating that 'intelligent-design theory' be taught in the public schools as balance for the 'holes' in evolutionary theory."[83] This apparently enraged the folks at the Discovery Institute, who blasted me with two consecutive blog posts for allegedly not recognizing that they had always "consistently opposed mandating intelligent design in public schools."[84] This is absurd, but perhaps it was part of their new strategy to outrun the stink of the *Kitzmiller* ruling.[85] On the advice of friends, I did not respond, but if I had it surely would have been worth pointing out that there is a difference between "mandating" and "advocating," and that if it were really true that they did not advocate teaching intelligent design in public schools, then what was the point of all of their amicus work on behalf of the defendants in the *Kitzmiller* case?

The picture painted here is a familiar one: pseudoscientists have no real understanding of or respect for what science entails. Furthermore, with the example of creationism/intelligent design, one also suspects that in this instance pseudoscience has bled over into denialism, where their proponents' minds are made up before they have even considered the evidence, because their views were not based on evidence in the first place. How then could intelligent design hope to pass itself off as a science? Part of the strategy is to try to exploit the weaknesses of science. Recall Sagan's criterion that science must be open to new ideas: Here the ID theorists complain that evolutionists are unfairly excluding their views from a fair hearing (even while they themselves refuse to acknowledge any evidence that contradicts their own views). "Teach the controversy" is their mantra. We need to examine all subjects impartially in science because ideas can come from anywhere. But if so, they complain, then why are the "scientific" claims of ID theory excluded from the biology classroom? Because there are none. I could spend many pages here dismembering the "scientific" evidence of ID theory point by point, but this has already been done brilliantly and at length by others.[86] If any readers feel the need to be convinced, I refer them to this work. For now, I am prepared to defer to Judge Jones's pithy conclusion in the *Kitzmiller* case that ID theory is "not science."[87]

Naturally, the ID theorist would push back and hope to engage in a lengthy debate about the origins of the eye and the missing link (too bad that evolutionists have an explanation for both).[88] But this raises the important question of whether such debates are even worth having—for do ID theorists even accept scientific standards of evidence? As pseudoscientists, ID theorists seek to insulate their views from refutation. They misunderstand the fundamental principle behind scientific reasoning, which is that one should form one's beliefs based on empirical evidence and commit to changing one's beliefs as new evidence comes forward.

And they have other misconceptions about science as well. For one, ID theorists often speak as if, in order to be taught in the science classroom, evolution by natural selection would have to be *completely proven* by the evidence. As we have seen, that is not how science works. As discussed in chapter 2, no matter how strong the evidence, scientific conclusions are never certain. Ah, the ID theorist may now object, isn't that why one should consider alternative theories? But the answer here is no, for any alternative theories are also bound by the principle of warrant based on sufficient evidence, and ID theory offers none. Note also that any "holes" in evolutionary theory would not necessarily suggest the validity of any particular alternative theory, unless it could explain them better.

Our conclusion? ID theory is just creationist ideology masquerading as science. The objection that evolution is not "settled science," and therefore that ID theory must at least be considered, is nonsense. One's place in the science curriculum must be *earned*. Of course, it is theoretically possible that evolutionary theory could be wrong, as any truly scientific theory could be. But this is overwhelmingly unlikely, as it is supported by a plethora of evidence from microbiology up through genetics. Merely to say, "Your theory could be wrong" or "Mine could be right" is not enough in science. One must offer some evidence. There must be some warrant. So even though it is theoretically true that evolution *could* be wrong, this does not somehow make the case for ID theory any more than it does for the parody religion of Pastafarianism and its "scientific" theory of the Flying Spaghetti Monster, which was invented as brilliant satire by an unemployed physics student during the *Kitzmiller* trial to illustrate the scientific bankruptcy of ID theory.[89] Indeed, if scientific uncertainty requires the acceptance of *all* alternative theories, must the astronomer also teach Flat Earth theory? Should we go back to caloric and phlogiston? In science, certainty may be unobtainable, but evidence is required.[90]

Another predictable complaint of ID theorists is that natural selection is "just a theory." Recall this myth about science from chapter 2. But to say that evolution is a theory does not dishonor it. To have a theory with a strong basis of evidential support—that backs up both its predictions and explanations, and is unified with other things that we believe in science—is a formidable thing. Is it any wonder that ID theory cannot keep up?

Yet let us now ask the provocative question: what if it could? What if there *were* some actual evidence in support of some prediction made by ID theory? Would we owe it a test then? I think we would. As noted earlier, even fringe claims must sometimes be taken seriously (which does not mean that they should be immediately inserted into the science classroom). This is because science *is* open to new ideas. In this way, some fringe theorists who bristle at the dismissal of their work as pseudoscience may have a legitimate complaint: where there is evidence to offer, science has no business dismissing an alternative theory based on anything *other* than evidence. But this means that if "pseudoscientific" theories are to be taken seriously, they must offer publicly available evidence that can be tested by others in the scientific community who do not already agree with the theory. So when they meet this standard, why don't scientists just go ahead and investigate?

Sometimes they do.

The Princeton Engineering Anomalies Research Lab

In 1979, Robert Jahn, the dean of the School of Engineering and Applied Science at Princeton University, opened a lab to "pursue rigorous scientific study of the interaction of human consciousness with sensitive physical devices, systems, and processes common to contemporary engineering practice."[91]

He wanted, in short, to study parapsychology. Dubbed the Princeton Engineering Anomalies Research (PEAR) lab, they spent the next twenty-eight years studying various effects, the most famous being psychokinesis, which is the alleged ability of the human mind to influence physical events.

A skeptic might immediately jump to the conclusion that this is pseudoscience, but remember that the proposal was to study this hypothesis in a scientific way. The PEAR team used random number generator (RNG) machines and asked their operators to try to influence them with their thoughts. And what they found was that there *was* a slight statistically significant effect of 0.00025. As Massimo Pigliucci puts it, although this is small, "if it were true it would still cause a revolution in the way we think of the basic behavior of matter and energy."[92]

What are we to make of this? First, it is important to remember our statistics:

Effect size is the size of the difference from random. Suppose you had a fair coin and flipped it 10,000 times and it came up heads 5,000 times. Then you painted the coin red and it came up heads 5,500 times. The 500 difference is the effect size.

Sample size is how many times you flipped the coin. If you flipped a painted coin only 20 times and it came up heads 11 of them, that is not very impressive. It could be due just to chance. But if you flip the coin 10,000 times and it came up heads 5,500 times, that is pretty impressive.

P-value is the probability that the effect you see is due to random chance. The reader will remember from our discussion in chapter 5 that p-value is not the same as effect size. The p-value is *influenced* by the effect size *but also by the sample size.* If you do your coin flips a large number of times and you still get a weird result, then there will be a lower p-value because it is unlikely to be due to chance. But effect size can also influence p-value. To get 5,500 out of 10,000 coin flips is actually a pretty big effect. With an effect that big, it is much less likely to be just due to randomness, so the p-value goes down.

Before we move on to the PEAR results, let's draw a couple of conclusions from the coin flip example. Remember that p-value does not tell you the cause of an effect, just the chance that you would see it if the null hypothesis were true. So it could be that the red paint you used on the coin was magic, or it could be that the weight distribution of the paint influenced the coin tosses to land on heads more often. One can't tell. What we can tell, however, is that having a larger number of trials means that even a very small effect size can be magnified. Suppose you had a perfectly normal looking unpainted coin, but it had a very small bias because of the way it was forged. As you flipped this coin one million times, the small bias would be magnified and would show up in the p-value. The effect size would still be small, but the p-value would go down because of the number of times you'd done the flip. Conclusion: it wasn't a fair coin. Similarly, a large effect size can have a dramatic effect on p-value even with a small number of flips. Suppose you took a fair coin and painted it red, and it came up heads ten times in a row. That is unlikely to be due to random chance. Maybe you used lead paint.

So what happened in the PEAR lab? They did the equivalent of flipping the coin for twenty-eight years in a row. The effect size was tiny, but the p-value was minuscule because of the number of trials. This shows that the effect could not have been due to random chance, right? Actually, no.

Although their hearts may have been in the right place—and I do not want to accuse the folks at the PEAR lab of any fraud or even bad intent—their results may have been due to a kind of massive unintentional p-hacking. In Pigliucci's discussion of PEAR research in his book *Nonsense on Stilts*, it becomes clear that the entire finding depends crucially on whether the random number generators were actually random.[93]

What evidence do we have that they were not? But this is the wrong question to ask. Remember that one cannot tell from a result what might have caused it, but before one embraces the extraordinary hypothesis that human minds can affect physical events, one must rule out all other plausible confounding factors.[94] Remember that once we painted the coins, we could not tell whether the effect was due to the "magic" properties of the paint or its differential weight on one side of the coin. Using Occam's razor, guess which one a skeptical scientist is going to embrace? Similarly, the effect at the PEAR lab could have been due either to psychokinesis or a faulty RNG. Until we rule out the latter, it is possible that all those years working with RNGs in the Princeton lab do not show that psychokinesis is possible at all, so much as they show that it is physically impossible to generate random numbers! As Robert Park puts it in *Voodoo Science*, "it is generally believed that there are no truly random machines. It may be, therefore, that the lack of randomness only begins to show up after many trials."[95]

How can we differentiate between these hypotheses? *Why* are there no random machines? This is an unanswered question, which goes to the heart of whether the psychokinetic hypothesis deserves further research (or indeed how it could even be tested). But in the meantime, it is important also to examine other methodological factors in the PEAR research. First, to their credit, they did ask other labs to try to replicate their results. These failed, but it is at least indicative of a good scientific attitude.[96] What about peer review? This is where it gets trickier. As the manager of the PEAR lab once put it, "We submitted our data for review to very good journals, but no one would review it. We have been very open with our data. But how do you get peer review when you don't have peers?"[97] What about controls? There is some indication that controls were implemented, but they were insufficient to meet the parameters demanded by PEAR's critics.

> Perhaps the most disconcerting thing about PEAR is the fact that suggestions by critics that should have been considered were routinely ignored. Physicist Bob Park reports, for example, that he suggested to Jahn two types of experiments that would have bypassed the main criticisms aimed at PEAR. Why not do a double-blind experiment? asked Park. Have a second RNG determine the task of the operator and do not let this determination be known to the one recording the results. This could have eliminated the charge of experimenter bias.[98]

While there has never been an allegation of fraud in PEAR research, it was at least suspicious that fully half of their effect size was due to the trials done by a single operator over all twenty-eight years, who was presumably an employee at the PEAR lab. Perhaps that individual merely had greater psychic abilities than the other operators. One may never know, for the PEAR lab closed for good in 2007. As Jahn put it,

> It's time for a new era, for someone else to figure out what the implications of our results are for human culture, for future study, and—if the findings are correct—what they say about our basic scientific attitude.[99]

Even while one may remain skeptical of PEAR's results, I find it cheering that the research was done in the first place and that it was taken seriously enough to critique it. While some called the lab an embarrassment to Princeton, I am not sure I can agree. The scientific attitude demands both rigor on the part of the researchers and an openness to consider scientifically produced data by the wider scientific community. Were these results scientific or were they pseudoscientific? I cannot bring myself to call them pseudoscience. I do not think that the researchers at PEAR were merely pretending to be scientists any more than those doing cold fusion or string theory. Perhaps they made mistakes in their methodology. Indeed, if it turns out to be the case that there actually is no such thing as a random number generator, perhaps the PEAR team should be credited with making this discovery! Then again, it would have been nice to see an elementary form of control, where they had let one RNG run all by itself (perhaps in another room) for twenty-eight years, with no attempt to influence it, and measure this against their experimental result. If the control machine showed 50 percent and the experimental one showed 50.00025, I would be more inclined to take the results seriously (both for showing that psychokinesis was possible and that RNGs were too).

Conclusion

Is it possible to think that one has the scientific attitude, but not really have it? Attitudes are funny things. Perhaps I am the only one who knows how I *feel* about using empirical evidence; only my private thoughts can tell me whether I am truly testing or insulating my theory. And even here I must respect the fact that there are numerous levels of self-awareness, complicated by the phenomenon of self-delusion. Yet the scientific attitude can also be measured through one's actions. If I profess to have the scientific attitude, then refuse to consider alternative evidence or make falsifiable predictions, I can be fairly criticized, whether I feel that my intentions are pure or not. The difference that marks off denialism and pseudoscience on one side and science on the other is more than just what is in the heart of the scientist or even the group of scientists who make up the scientific community. It is also in the actions taken by the individual scientist, and the members of his or her profession, who make good on the idea that science really does care about empirical evidence. As elsewhere in life, we measure attitudes not just by thought but also by behavior.

9 The Case for the Social Sciences

In the last two chapters, we witnessed several examples of failure to emulate the scientific attitude. Whether because of fraud, denialism, or pseudoscience, many who purport to care about evidence fail to live up to the highest standards of empirical inquiry. One key issue to consider here is motivation. Some of these failures occur because the people involved are not really aspiring to be scientific (perhaps because they care about something like ideology, ego, or money ahead of scientific integrity) and just want a shortcut to scientific glory.

But what about those people who work in fields that *do* want to become more scientific—and are willing to work hard for it—but just do not fully appreciate the role that the scientific attitude might have in getting them there? In chapter 6, we saw the power of the scientific attitude to transform a previously unscientific field into the science of medicine. Is the same path now open to the social sciences? For years, many have argued that if the social sciences (economics, psychology, sociology, anthropology, history, and political science) could emulate the "scientific method" of the natural sciences, they too could become more scientific. But this simple advice faces several problems.

Challenges for a Science of Human Behavior

There are many different ways of conducting social inquiry. Social psychologists have found it convenient to rely on controlled experiments (and behavioral economists are just beginning to catch up with them), but in other areas of social investigation it is just not possible to run the data twice.[1] In sociology, we have case studies. In anthropology, we have fieldwork. And, up until recently, neoclassical economists disdained the idea that reliance on simplifying assumptions undercut the applicability of their theoretical models to human behavior. Of course, in this way, the social sciences are not entirely different from the natural sciences. With Newtonian physics as a model, it is easy to overlook the fact that the study of nature too has its methodological diversity: geologists cannot run controlled experiments, and biologists are often prevented from making precise predictions. Nonetheless, ever since the Logical Positivists (and Popper) claimed that what was special about science was its method, many have felt that the social sciences would be better off if they tried to model themselves on the type of inquiry that goes on in the natural sciences.

This idea has caused some push back over the years from both social scientists and philosophers of social science, who have held that one cannot hope to do social science in the same way as natural science. The subject matter is just too different. What we want to know about human behavior is often at odds with the desire merely to reduce actions to their causal forces. So if it were true that there is really only one way to do science—one that is defined by the unique methodology of natural science (if not "scientific method" itself)—it would be easy to see why some might lose hope in the idea that social science could become more scientific.

In earlier work, I spent a good deal of effort trying to identify flaws in those arguments that held there was a fundamental barrier to having a science of human action because of the complexity or openness of its subject matter, the inability to perform controlled social experiments, and the special problems created by subjectivity and free will in social inquiry.[2] I continue to believe that my arguments here are valid, primarily because complexity and openness are also a part of natural scientific inquiry, and the other alleged barriers have less effect on the actual performance of social inquiry than their critics might assume. I also believe that these problems are overblown; if they were truly a barrier to inquiry they would likely show that much of natural science should not work either. But for the time being, I wish to focus on a more promising path for defending the possibility of a science of human behavior, for I now realize that what has been missing from the social sciences all these years is not proper method, but the right attitude toward empirical evidence.[3]

Like Popper, I have never believed that there is such a thing as scientific method, but I have held that what makes science special is its methodology.[4] Popper, of course, famously held that the social sciences were not falsifiable, so they could not be scientific.[5] I have argued against this and held that in every meaningful way there could be methodological parity between natural and social inquiry. That may be, but it overlooks one crucial point. What makes science special—both natural and social—is not just the way that scientists conduct their inquiry, but the attitude that informs their practices.

Too much social research is embarrassingly unrigorous. Not only are the methods sometimes poor, much more damning is the nonempirical attitude that lies behind them. Many allegedly scientific studies on immigration, guns, the death penalty, and other important social topics are infected by their investigator's political or ideological views, so that it is all but expected that some researchers will discover results that are squarely in line with liberal political beliefs, while others will produce conservative results that are directly opposed to them. A good example here is the question of whether immigrants "pay their own way" or are a "net drag" on the American economy. If this is truly an empirical question (and I think that it is) then why can I cite five studies that show that immigrants are a net plus, five more that show that they are not, and can probably predict which studies came out of which research centers and by whom they were written?[6] I am not here accusing anyone of fraud or pseudoscience. These are purportedly rigorous social scientific studies performed by well-respected scholars—it is just that their findings *about factual matters* flatly contradict one another. This would not be tolerated in physics, so why is it tolerated in sociology? Is it any wonder that politicians in Washington, DC, are so skeptical about basing their policies on social scientific work and instead cherry pick their favorite studies to support their preferred ideology?

The truth is that such questions *are* open to empirical study and it *is* possible for social science to study them scientifically. There are right and wrong answers to our questions about human behavior. Do humans experience a "backfire effect" when exposed to evidence that contradicts their opinion on an empirical (rather than a normative) question, such as whether there were weapons of mass destruction in Iraq or President George W. Bush proposed a complete ban on stem cell research? Is there such a thing as implicit bias, and if so, how can it be measured? Such questions can be, and have been, studied scientifically.[7] Although social scientists may continue to disagree (and indeed, this is a healthy sign in ongoing research), their disagreements should focus on the best way to investigate these questions, not whether the answers produced are politically acceptable. Having the scientific attitude toward evidence is just as necessary in the study of human behavior as it is in the study of nature.

Yet there are numerous problems in contemporary social scientific research:

(1) *Too much theory*: A number of social scientific studies propose answers that have not been tested against evidence. The classic example here is neoclassical economics, where a number of simplifying assumptions—perfect rationality, perfect information—resulted in beautiful quantitative models that had little to do with actual human behavior.

(2) *Lack of experimentation/data*: Except for social psychology and the newly emerging field of behavioral economics, much of social science still does not rely on experimentation, even where it is possible. For example, it is sometimes offered as justification for putting sex offenders on a public database that doing so reduces the recidivism rate. This must be measured, though, against what the recidivism rate *would have been* absent the Sex Offender Registry Board (SORB), which is difficult to measure and has produced varying answers.[8] This exacerbates the difficulty in (1), whereby favored theoretical explanations are accepted even when they have not been tested against any experimental evidence.

(3) *Fuzzy concepts*: In some social scientific studies, the results are questionable because of the use of "proxy" concepts for what one really wishes to measure. Recent examples include measuring "warmth" as a proxy for "trustworthiness" (see details in next section), which can lead to misleading conclusions.

(4) *Ideological infection*: This problem is rampant throughout the social sciences, especially on topics that are politically charged. One recent example is the bastardization of empirical work on the deterrence effect of capital punishment or the effectiveness of gun control on mitigating crime. If one knows in advance what one wants to find, one will likely find it.[9]

(5) *Cherry picking*: As we've seen, the use of statistics allows multiple "degrees of freedom" to scientific researchers, but this is the most likely to be abused. In studies on immigration, for instance, a great deal of the difference between them is a result of alternative ways of counting the "costs" incurred by immigration. This is obviously also related to (4) above. If we know our conclusion, we may shop for the data to support it.

(6) *Lack of data sharing*: As Trivers reports, there are numerous documented cases of researchers failing to share their data in psychological studies, despite a requirement from APA-sponsored journals to do so.[10] When data were later analyzed, errors were found most commonly in the direction of the researcher's hypothesis.

(7) *Lack of replication*: As previously discussed, psychology is undergoing a reproducibility crisis. One might validly argue that the initial finding that nearly two-thirds of psychology studies were irreproducible was overblown (see Gilbert et al. 2016), but it is nonetheless shocking that most studies *are not even attempted to be replicated*. This can lead to difficulties, where errors can sneak through.

(8) *Questionable causation*: It is gospel in statistical research that "correlation does not equal causation," yet some social scientific studies continue to highlight provocative results of questionable value. One recent sociological study, for instance, found that matriculating at a selective college was correlated with parental visitation at art museums, *without explicitly suggesting that this was likely an artifact of parental income*.[11]

All of these problems can also be found to some degree in natural scientific work. Some of the other problems identified in earlier chapters (p-hacking, problems with peer review) can be present in social scientific work as well. The issue is not so much that these difficulties are *unique* to social inquiry, as that some of them are *especially prevalent* there. Even if natural science suffers from them as well, the problems for social science are proportionally greater.

The problem with social science is not that it is failing to follow some prescribed method, or even to embrace certain scientific procedures, but that a number of its practices are not yet instantiated at the level of group practice so that they demonstrate a discipline-wide commitment to the scientific attitude. Fuzzy concepts or errors from questionable causation may not be such a big problem if one can count on one's colleagues to catch them, but in many instances—in an environment in which data are not shared and replication is not the norm—too many errors slip by. Social science no less than natural science needs to embrace the scientific attitude toward evidence and realize that the only way to settle an empirical dispute is with empirical evidence. It should be recognized as the embarrassment that it is that there is so much opinion, intuition, theory, and ideology in social scientific research. Just as we now look back on bare-handed surgery and cringe, may we someday ask "Why didn't someone test that hypothesis?" in social science? Nothing keeps people honest like public scrutiny. We need to do more data sharing and replication in the social sciences. We need better peer review and real scientific controls. And we need to recognize that it is shameful that up until recently much of social science has not even made an attempt to be experimental. Compared to the old neoclassical model in economics, the new behavioral one is a breath of fresh air. And this is all made possible by embracing the scientific attitude.

If social scientists were more committed at both the individual and group level to reliance on evidence and constructing better procedures for capitalizing on this, the social sciences would be better off. In this, they can follow the same path that was earlier laid out by medicine. Whether one is performing a clinical trial or doing fieldwork, the appropriate attitude to have in social inquiry should be that previously noted by Emile Durkheim: "When [we] penetrate the social world… [we] must be prepared for discoveries which will surprise and disturb [us]."[12] We must abandon the belief that just because we are human we already basically understand how human behavior works. Where possible, we need to perform experiments that challenge our preconceptions, so that we can discover how human action actually works, rather than how our mathematical models and high theory tell us it should.

And all of this applies equally to qualitative work as it does to quantitative. While it is true that in social science there is some evidence that may be irreducibly qualitative (see Clifford Geertz's work on "thick description" in his book *The Interpretation of Cultures* [Basic Books, 1973]), one must still be concerned with how to measure it. Indeed, in the case of qualitative work, we must be especially on guard against hubris and bias. Our intuitions about human nature are likely no more profound than eighteenth-century physicians' were about infection.[13] The data can and should surprise us. Just because a result "feels right" does not mean it is accurate. Cognitive bias and all of the other threats to good scientific work in the study of nature are no less a threat to those who study human behavior. The revolution in social science may be attitudinal rather than methodological, but this does not mean that it should not reach the four corners of how we have been doing our inquiry.

For years, many have thought that one could improve social science by making it more "objective." The Logical Positivists in particular held fast to the fact–value distinction, which said that scientists should be concerned only with their results and not with how these might be used. But they were wrong. While it is true that we should not let our hopes, wishes, beliefs, and "values" color our inquiry into the "facts" about human behavior, this does not mean that values are unimportant. Indeed, as it turns out, our commitment to the scientific attitude is an essential value in conducting scientific inquiry. The key to having a more rigorous social science is not scientific method, but the scientific attitude.

A Way Forward: Emulating Medicine

If we think back to the state of medicine at the time of Semmelweis, the analogies with social science are compelling. Knowledge and procedures were based on folk wisdom, intuition, and custom. Experiments were few. When someone had a theory, it was thought to be enough to consider whether it "made sense," even if there was no empirical evidence in its favor. Indeed, the very idea of attempting to gather evidence to test a theory flew in the face of belief that medical practitioners *already knew* what was behind most illnesses. Despite the shocking ignorance and backward practices of medicine throughout most of its history, theories were abundant and ideas were rarely challenged or put to the test. Indeed, this is what was so revolutionary about Semmelweis. He wanted to know whether his ideas held up by testing them in practice. He understood that knowledge accumulates as incorrect hypotheses are eliminated based on lack of fit with the evidence. Yet his approach was wholeheartedly resisted by virtually all of his colleagues.

Medicine at the time did not yet have the scientific attitude. Does social science now? In some cases it does, but the problem is that even in those instances where good work is being done many feel free to ignore it. In a society in which law enforcement continues to rely on eyewitness testimony and perform nonsequential criminal lineups, despite the alarmingly high rate of false positives with these methods, we have to wonder whether this is just another instance where practice trails theory.[14] Public policy on crime, the death penalty, immigration, and gun control is rarely based on actual empirical study. Yet at least part of the problem also seems due to the inconsistent standards that have always dogged social scientific research. This has handicapped the reputation of the social sciences and made it difficult for good work to get noticed. As we have seen, when so many studies fail to be replicated, or draw different conclusions from the same set of facts, it does not instill confidence. Whether this is because of sloppy methodology, ideological infection, or other problems, the result is that even if there are right and wrong answers to many of our questions about human action, most social scientists are not yet in a position to find them. It is not that *none* of the work in social science is rigorous enough, but when policy makers (and sometimes even other researchers) are not sure which results are reliable, it drives down the status of the entire field.

Medicine was once held in similarly low repute, but it broke out of its prescientific "dark ages" because of individual breakthroughs that became the standard for group practice and some degree of standardization of what counted as evidence. To date, the social sciences have yet to complete their evidence-based revolution. We can find some examples today of the scientific attitude at work in social inquiry that have enjoyed some success, but there has not yet been a discipline-wide acceptance of the notion that the study of human behavior needs to be based on theories and explanations that are relentlessly tested against what we have learned through experiment and observation. As in prescientific medicine, too much of today's social science relies on ideology, hunches, and intuition.

In the next section, I will provide an example of what a social science that fully embraced the scientific attitude might look like. Before we get there, however, it is important to consider one remaining issue that many have felt to be an insuperable barrier to the pursuit of a science of human behavior. Some have said that social science is unique because of the inherent problem of humans studying other humans—that our values will inevitably interfere with any "objective" empirical inquiry. This is the problem of *subjectivity bias*. Yet it is important to remember that in medicine we have an example of a field that has already solved this problem and moved forward as a science.

In its subject matter, medicine is in many ways like social science. We have irreducible values that will inevitably guide our inquiry: we value life over death, health over disease. We cannot even begin to embrace the "disinterested" pose of the scientist who does not care about his or her inquiry beyond finding the right answer. Medical scientists desperately hope that some theories will work because lives hang in the balance. But how do they deal with this? Not by throwing up their hands and admitting defeat, but rather by relying on good scientific practices like randomized double-blind clinical trials, peer review, and disclosure of conflicts of interest. The placebo effect is real, for both patients and their doctors. If we want a medicine to work, we might subtly influence the patient to think that it does. But whom would this serve? When dealing with factual matters, medical researchers realize that influencing their results through their own expectations is nearly as bad as fudging them. So they guard against the hubris of thinking that they already know the answer by instituting methodological safeguards. They protect what they care about by recognizing the danger of bias.

The mere presence of values or caring about what you study does not undercut the possibility of science. We can still learn from experience, even if we are fully invested in hoping that one medicine will work or one theory is true, as long as we do not let this get in the way of good scientific practice. We can still have the scientific attitude, even in the presence of other values that may exist alongside it. Indeed, it is precisely *because* medical researchers and physicians recognize that they may be biased that they have instituted the sorts of practices that are consonant with the scientific attitude. They do not wish to stop caring about human life, they merely want to do better science so that they can promote health over disease. In fact, if we truly care about human outcomes, it is *better* to learn from experience, as the history of medicine so clearly demonstrates. It is only when we take steps to preserve our objectivity—instead of pretending that this is not necessary or that it is impossible—that we can do better science.

Like medicine, social science is subjective. And it is also normative. We have a stake not just in knowing how things are but also in using this knowledge to make things the way we think they should be. We study voting behavior in the interest of preserving democratic values. We study the relationship between inflation and unemployment in order to mitigate the next recession. Yet unlike medicine, so far social scientists have not proven to be very effective in finding a way to wall off positive inquiry from normative expectations, which leads to the problem that instead of acquiring objective knowledge we may only be indulging in confirmation bias and wishful thinking. This is the real barrier to a better social science. It is not just that we have ineffective tools or a recalcitrant subject matter; it is that at some level *we do not yet have enough respect for our own ignorance to keep ourselves honest by comparing our ideas relentlessly against the data*. The challenge in social science is to find a way to preserve our values without letting them interfere with empirical investigation. We need to understand the world before we can change it. In medicine, the answer was controlled experimentation. What might it be in social science?

Examples of Good and Bad Social Science

Even when social scientists do "research," it is often not experimental. This means that a good deal of what passes for social scientific "evidence" is based on extrapolations from surveys and other data sets that may have been conducted by other researchers for other purposes. But this can lead to various methodological problems such as confusion between causation and correlation, the use of fuzzy concepts, and some of the other weaknesses we spoke about earlier in this chapter. It is one thing to say that "bad" social science is all theory and no evidence, infected with ideology, does not rely enough on actual experimentation, is not replicable, and so on, but it is another to see this in action.

One example of poorly conducted social scientific research can be found in a 2013 article by Susan Fiske and Cydney Dupree entitled "Gaining Trust as Well as Respect in Communicating to Motivated Audiences about Science Topics," which was published in the Perspectives section of the *Proceedings of the National Academy of Science.*[15] In this study, the researchers set out to study an issue that has great importance for the defense of science: whether the allegedly low trustworthiness of scientists may be undermining their persuasiveness on factual questions such as climate change. Does it come as a surprise that scientists are seen as untrustworthy? Fiske and Dupree purport to have empirical evidence for this.

In their study, the researchers first conducted an online poll of American adults to ask them to list typical American jobs. The researchers then chose the most commonly mentioned forty-two jobs, which included scientists, researchers, professors, and teachers.[16] In the next step, they polled a new sample to ask about the "warmth" versus "competence" of practitioners of these professions. Here it was found that scientists rated highly on expertise (competence) but relatively low on warmth (trustworthiness). What does warmth have to do with trustworthiness? Their hypothesis was that trustworthiness is positively correlated with warmth and friendliness. In short, if someone is judged to be "on my side" then that person is more likely to be trusted. But whereas there is empirical work to show that if someone is judged to be "like us" we are more likely to trust that person,[17] it is a great leap to then start using "warmth" and "trustworthiness" as interchangeable proxies for one another.

First, one should pay attention to the leap from saying (1) "if *X* is on my side, then *X* is more trustworthy" to saying (2) "if *X* is *not* on my side, then *X* is *less* trustworthy." By elementary logic, we understand that statement (2) is not implied by statement (1), nor vice versa. Indeed, the leap from (1) to (2) is the classic logical error of denying the antecedent. This means that even if there were empirical evidence in support of the truth of statement (1), the truth of statement (2) is still in question. Nowhere in Fiske and Dupree's article do they cite any evidence in support of statement (2), yet the biconditional link between "being on my side" and "being trustworthy" is the crux of their conclusion that it is methodologically sound to use "warmth" as a proxy to measure "trustworthiness."[18] Isn't it conceivable that scientists could be judged as not warm yet nonetheless trustworthy? Indeed, wouldn't it have been more direct if the researchers had simply asked their subjects to rate the trustworthiness of various professions? One wonders what the result might have been. For whatever reason, however, the researchers chose not to take this route and instead skip blithely back and forth between measurements of warmth and conclusions about trust throughout their article.

> [Scientists] earn respect but not trust. Being seen as competent but cold might not seem problematic until one recalls that communicator credibility requires not just status and expertise (competence) but also trustworthiness (warmth).... Even if scientists are respected as competent, they may not be trusted as warm.[19]

This is a classic example of the use of fuzzy concepts in social scientific research, where dissimilar concepts are treated as interchangeable, presumably because one of them is easier to measure than the other. In this case, I am not convinced of that, because "trust" is hardly an esoteric concept that would be unreportable by research subjects, but we nonetheless find in this article a conclusion that scientists have a "trust" problem rather than a "warmth" problem, based on zero direct measurement of the concept of trust itself.[20]

This is unfortunate, because the researchers' own study would seem to give reason to doubt the truth of their own conclusions. In a follow up, Fiske and Dupree report that as a final step they singled out climate scientists for further review, and here polled a fresh sample of subjects with a slightly different methodology for the measurement of trust. Here, instead of allegedly measuring "trust," they instead sought to measure "distrust" through the use of a seven-item scale that included things like perceptions of "motive to lie with statistics, complicate a simple story, show superiority, gain research money, pursue a liberal agenda, provoke the public, and hurt big corporations."[21] The researchers were surprised to find that climate scientists were judged to be more trustworthy than scientists in general (measured against their previous poll). What might be the reason for this? They offer the hypothesis that the scale was different (which raises the question of why they made the decision to use a different scale), but also float the idea that climate scientists perhaps had a more "constructive approach to the public, balancing expertise (competence) with trustworthiness (warmth), together facilitating communicator credibility."[22] I find this to be a questionable conclusion, for in the final part of the study there was no measurement at all of the "warmth" of climate scientists, yet the researchers once again feel comfortable drawing parallels between *trustworthiness* and *warmth*.[23]

By way of contrast, I will now explore an example of good social scientific work that is based firmly in the scientific attitude, uses empirical evidence to challenge an intuitive theoretical hypothesis, and employs experimental methods to measure human motivation directly through human action. In Sheena Iyengar's work on the paradox of choice, we face a classic social scientific dilemma. How can something as amorphous as human motivation be measured through empirical evidence? According to neoclassical economics, we measure consumer desire directly through marketplace behavior. People will buy what they want, and the price is a reflection of how much the good is valued. To work out the mathematical details, however, a few "simplifying assumptions" are required. First, we assume that our preferences are rational. If I like cherry pie more than apple, and apple more than blueberry, it is assumed that I like cherry more than blueberry.[24] Second, we assume that consumers have perfect information about prices. Although this is widely known to be untrue in individual cases, it is a core assumption of neoclassical economics, for it is needed to explain how it is that the market as a whole performs the magical task of ordering preferences through prices.[25] Although it is acknowledged that actual consumers may make "mistakes" in the marketplace (for instance, they did not know that cherry pie was on sale at a nearby market), the model purports to work because if they had known this, they would have changed their behavior. Finally, the neoclassical model assumes that "more is better." This is not to say that there is no such thing as diminishing marginal utility—that last bite of cherry pie probably does not taste as good as the first one—but it is to say that for consumers it is better to have more choices in the marketplace, for this is how one's preferences can be maximized.

In Sheena Iyengar's work, she sought to test this last assumption directly through experiment. The stakes were high, for if she could show that this simplifying assumption was wrong, then, together with Herbert Simon's earlier work undermining "perfect information," the neoclassical model may be in jeopardy. Iyengar and her colleague Mark Lepper set up a controlled consumer choice experiment in a grocery store where shoppers were offered the chance to taste different kinds of jam. In the control condition, shoppers were offered twenty-four different choices. In the experimental condition, this was decreased to six options. To ensure that different shoppers were present for the two conditions, the displays were rotated every two hours and other scientific controls were put in place. Iyengar and Lepper sought to measure two things: (1) how many different flavors of jam the shoppers chose to taste and (2) how much total jam they actually bought when they checked out of the store. To measure the latter, everyone who stopped by to taste was given a coded coupon, so that the experimenters could track whether the number of jams in the display affected later purchasing behavior. And did it ever. Even though the initial display of twenty-four jams attracted slightly more customer interest, their later purchasing behavior was quite low when measured against those who had visited the booth with only six jams. Although each display attracted an equal number of jam tasters (thus removing the fact of tasting as a causal variable to explain the difference), the shoppers who had visited the display with twenty-four jams used their coupons only 3 percent of the time, whereas those who visited the display with only six jams used theirs 30 percent of the time.

What might account for this? In their analysis, Iyengar and Lepper speculated that the shoppers might have been overwhelmed in the first condition.[26] Even when they tasted a few jams, this was such a small percentage of the total display that they perhaps felt they could not be sure they had chosen the best one, so they chose not to buy any at all. In the second condition, however, shoppers might have been better able to rationalize making a choice based on a proportionally larger sampling. As it turned out, people wanted fewer choices. Although they might not have realized it, their own behavior revealed a surprising fact about human motivation.[27]

Although this may sound like a trivial experiment, the implications are far reaching. One of the most important direct applications of Iyengar and Lepper's finding was to the problem of undersaving in 401k plans, where new employees are customarily overwhelmed by the number of options for investing their money and so choose to put off the decision, which effectively means choosing not to invest any money at all. In *Respecting Truth*, I have explored a number of other implications of this research ranging from automatic enrollment in retirement plans to the introduction of "target date" retirement funds.[28] Not only is this good social science, but its positive impact on human lives has been considerable.

For present purposes, the point is this. Even in a situation where we may feel most in touch with our subject matter—human preference and desire—we can be wrong about what influences our behavior. If you ask people whether they want more or fewer choices, most will say they want more. But their actual behavior belies this. The results of experimental evidence in the study of human action can surprise us. Even concepts as seemingly qualitative as desire, motivation, and human choice can be measured by experimentation rather than mere intuition, theory, or verbal report.

Here again we are reminded of Semmelweis. How do we know before we have conducted an experiment what is true? Our intuitions may feel solid, but experiment shows that they can fail us. And this is as true in social science as it is in medicine. Having the facts about human behavior can be just as useful in public policy as in the diagnosis and treatment of human disease. Thus the scientific attitude is to be recommended just as heartily in social science as it is in any empirical subject. If we care about evidence and are willing to change our minds about a theory based on evidence, what better example might we have before us than the success of Iyengar and Lepper's experiment? Just as the elegance of Pasteur's experimental model allowed him to overthrow the outdated idea of spontaneous generation, could economics now move forward owing to recognition of the impact of cognitive bias and irrationality on human choice?

And perhaps this same approach might work throughout the social sciences. All of the recent work on cognitive bias, for instance, might help us to develop a more effective approach to science education and the correction of public misperceptions about climate change. If the researchers cited by Fiske and Dupree as the foundation for their work are right (which has nothing to do with the question of any purported connection between warmth and trustworthiness), then *attitude* is as much a part of making up our mind as *evidence*.

> First, scientists may misunderstand the sources of lay beliefs. People are no idiots. The public's issue with science is not necessarily ignorance. The public increasingly knows more than before about climate change's causes.... Potential divides between scientists and the public are not merely about sheer knowledge in any simple way.
>
> The second, often-neglected factor is the other side of attitudes. Attitudes are evaluations that include both cognition (beliefs) and affect (feelings, emotions). Acting on attitudes involves both cognitive capacity and motivation. Attitudes show an intrinsic pressure for consistency between cognition and affect, so for most attitudes, both are relevant. When attitudes do tilt toward emphasizing either cognition or affect, persuasion is more effective when it matches the type of attitude. In the domain of climate change, for example, affect and values together motivate climate cognition. If public attitudes have two sides—belief and affect—what is their role in scientific communication?[29]

If this is true, what breakthroughs might be possible once we gain better experimental evidence for how the human mind really works? Actual human beings do not have perfect information; neither are they perfectly rational. We know that our reason is buffeted by built-in cognitive biases, which allow a sea of emotions, misperceptions, and desires to cloud our reasoning. If we seek to do a better job of convincing people to accept the scientific consensus on a topic like climate change, this provides a valuable incentive for social scientists to get their own house in order. Once they have found a way to make their own discipline more scientific, perhaps they can play a more active role in helping us to defend the enterprise of science as a whole.

10 Valuing Science

If the conception of science that I have outlined in this book is correct, we are now in a position to do three things: (1) understand and (2) defend the success of science where it has heretofore existed, and (3) grow it elsewhere. If fields like the social sciences wish to become more rigorous, they must follow the path traveled by others fields such as medicine: they must embrace the scientific attitude.

The scientific attitude has helped us to realize that what is most special about science is not the alleged "scientific method" that it follows, but rather its respect for the power of empirical evidence to shape and change our theories, and reliance on the practices of critical scrutiny by our peers to catch our mistakes if we do not. Evidence matters in science, and recognition of this is the most important value that marks the difference between those who practice science and those who do not. Even if evidence cannot uniquely determine which theory is "true," deference to empirical evidence is what gives science its special explanatory power. It keeps us on track even when our imperfect ideas and human weaknesses threaten to throw us off course.

We also now understand why ideological theories like intelligent design and denialism about climate change should not be considered scientific, for they in some ways rely on the antithesis of the scientific attitude. They champion ideology over evidence. They have no humility over the fact that scientific investigation pushes us not toward certainty, but more likely toward abandoning some false idea that we desperately wanted to believe. There is more to scientific explanation than merely "getting it right" with technology or attempting to prove a theory.

But scientific explanation is no less special if we finally accept it for what it is. We do not need to pretend that every scientific theory is in principle reducible to physics, or that the success of scientific theories means that they are more probably true. Science is a rational process by which we learn how to constantly reevaluate and discard our prejudices, wishes, and hunches about the world and replace them with conclusions that can be squared with the data of human experience. This is the root of scientific warrant. Despite the fact that science can never get us to "truth," it is still a unique and remarkable way of knowing. Finally, to realize that science is about one's *attitude* and not one's *method*—about seeking justification rather than jumping to a conclusion about truth—is nonetheless to have a uniquely powerful tool in our explanatory arsenal. For all its limitations, I believe that science is the greatest invention the human mind has ever created for gathering empirical knowledge. As such, it is worth understanding, emulating, and defending.

There remains, however, a pregnant question, for now that I have suggested such a robust defense of science, some may wonder whether I should take one more swing at developing the scientific attitude into a new criterion of demarcation. I trust that I have made clear in earlier chapters why I do not think that this is a wise (or necessary) move. Now that we are nearer the end, however, let me say a few more words about this. Throughout this book I have left the definition of the scientific attitude somewhat vague and refused to list too many criteria, because I think that science is ill-served by trying to specify too discretely what counts as science and what does not. The temptation here may be to try to reduce the scientific attitude to some sort of methodological formula, but my hope is that the scientific attitude will be used instead not to build a wall, but to light a path by showing what other disciplines must do if they wish to become more scientific.

There is a difference between saying that the scientific attitude is a necessary condition and saying that it is a sufficient one.[1] As stated earlier, to the extent that we are exploring and defending what is special about science, I believe that we should give up on the use of sufficient conditions. There are many ways for an investigation to fail to be science. This is why I have deliberately proposed the scientific attitude only as a necessary condition. But my hesitation in the face of playing the "if and only if" game, which earlier philosophers of science have said was required to demarcate science from nonscience, helps us to avoid a much worse pitfall—for if we pin our hopes on finding a set of necessary and sufficient conditions for science, the standards may be *so* high that we will never find a satisfactory answer to the question of what is distinctive about science, which leaves science undefended.[2] Some will point out that I too have a flank open—that perhaps my account allows *too much* to count as science. If the scientific attitude is all that is necessary, why not count every instance in which someone uses the process of elimination or relies on observational evidence as science? Remember our earlier dust up over the question of whether the act of looking for one's keys should count as science? Remember the farce of Bode's law?

Perhaps there is further work to be done to nail down any remaining necessary conditions for doing science. Having a full-blown theory may be one of them. If the scientific attitude requires that one must use *evidence* to test a *theory*, we had better be sure that we have a theory to test! As Bode's law demonstrates, science is more than a series of lucky guesses. If our "theory" fits the evidence only by accident, this provides no warrant. Newtonian theory may get us to the Moon and back (even if it is wrong), while Bode's law successfully predicted two new planets (even though it was based on no theory whatsoever), yet the first has warrant and the second does not. Why? Because instrumental success is not the sum total of what is distinctive about science. We are also concerned with justification. When we use the scientific attitude, we are testing a theory against the evidence. Though a theory may turn out to be wrong, this does not mean it is not scientific. Yet with Bode's law there was no theory to test. The success of Bode's law shows that even in empirical inquiry someone can get lucky.[3] But this is not science. Success alone does not vindicate science. Just as the falsity of a theory does not make it unscientific, neither does mere fit with the evidence make a hypothesis scientific. There has to be a theory at stake. Which means that when we say that the scientific attitude requires us to *care about evidence and use it to modify our theory*, we had better take seriously the idea not just that science relies on *evidence*, but also that it requires a *theory* to be warranted on its basis. This is the backbone of scientific explanation.

If scientific theories can never be proven true, we may never be certain how the world works. But this does not mean there is a better path. Even if the goal of science cannot be to get us to certainty (because this would outstrip even the greatest warrant we can infer from fit between a theory and empirical evidence) this is not a threat to the epistemological authority of science. Despite its limitations, science still provides a better way of knowing the empirical world than any of its competitors, because it is based on the founding assumption of science: that we come to know the world best through the evidence that it provides. We cannot rely on our intuition or our guesses. And we also cannot pretend that the empirical success of our theories lands us closer to truth. Yet there is simply no better way to understand the empirical world than to measure our hypotheses against the data of experience.

For decades there has been a feud within the philosophy of science between those who think that science is special (and customarily try to situate that privilege within its allegedly unique methodology) and those who think that the process of scientific reasoning is no better or worse than any other human endeavor (owing to the social factors, attitudes, and interests that permeate all human conduct). What irony, therefore, if it turned out that what is most distinctive about science is the attitudes and values that scientists hold toward their inquiry!

At bottom, what is special about science turns out not to be its method. In trying to explain science (and also what other fields should emulate if they wish to become more scientific), we must focus on scientific values. Previous accounts of science have spent a good deal of time arguing about methodology, both in trying to see whether we can demarcate science from nonscience, and in probing whether, if fields like the social sciences could just undergo methodological reform, they too could become more scientific. But this turns out to be the wrong focus.

Throughout this book, I have argued that it is of limited value to spend most of our time in the philosophy of science studying the methods of those successful fields like astronomy and physics that have already become sciences. While there is a lot to learn from them, this will not get us to the deepest level of understanding what is distinctive about science. Instead, what I have done is look at those fields that are either failing to become sciences (because they are hopelessly ideological, have the wrong values, or are practiced by people who simply don't care about formulating their beliefs based on evidence) or those that are still hoping to become scientific (like the social sciences) and are struggling to do better. An examination of what is missing in these fields reveals what is necessary for science.

The scientific attitude is elusive, but it has been there the whole time. It was there in Bacon's "virtues" and Popper's account of falsification. It was there in Kuhn's account of paradigm shifts. And it has been there even in the critics who have said that there is nothing special about the methodology of science. As it turns out, there may be no hard and fast criterion of demarcation between science and nonscience. Yet science is nonetheless real and distinctive, and we ought to pay proper respect to it as a privileged way of knowing. If we care about empirical evidence and use it to shape our theories, we are on the road to science. If not, we will remain mired in the ditch of ideology, superstition, and confusion.

It is hard to hit a target one does not aim at. One of the most important things about science is that it aims at an understanding that, although it may not get us to truth, at least shapes the imaginative ideal of empirical inquiry as something that is noble and worth emulating. Science is one of the few activities where we catch humanity at its best; despite any selfish motives or petty agendas, we can cut through all this by caring about the right things, critiquing one another's work, and never forgetting the common goal toward which we all aim: to know something from nature (or from human experience) that we did not know before.

But just as science is real, the scientific attitude is real as well. Even if it is difficult to define and hard to measure, it makes up the core difference between what is science and what is not. To some this may seem surprising and hard to accept. It would be nice to have a "harder" criterion. Something logical perhaps; something we could use to cleave the world of inquiry neatly in two. This has long been the dream of the philosophy of science. But I do not think this dream can be fulfilled. This does not mean, however, that we cannot defend or emulate science, now that we are awake to its true nature. Science is still special, even if it is not quite what we thought it was.

Notes

Introduction

1. P. H. Gleick, R. M. Adams, R. M. Amasino, E. Anders, D. J. Anderson, W. W. Anderson, et al., "Climate Change and the Integrity of Science," *Science* 328, no. 5979 (2010): 689–690, http://science.sciencemag.org/content/328/5979/689.

2. "On Energy Policy, Romney's Emphasis Has Shifted," NPR, April 2, 2012, http://www.npr.org/2012/04/02/149812295/on-energy-policy-another-shift-for-romney.

3. "Scientific Evidence Doesn't Support Global Warming, Sen. Ted Cruz Says," NPR, Dec. 9, 2015, http://www.npr.org/2015/12/09/459026242/scientific-evidence-doesn-t-support-global-warming-sen-ted-cruz-says.

4. Oliver Milman, "Trump to Scrap NASA Climate Research in Crackdown on 'Politicized Science,'" *Guardian*, Nov. 23, 2016, https://www.theguardian.com/environment/2016/nov/22/nasa-earth-donald-trump-eliminate-climate-change-research.

5. Recent examples of disparaging remarks about philosophy by such renowned scientists as Stephen Hawking, Lawrence Krauss, and Neil deGrasse Tyson are discussed in Massimo Pigliucci's essay "Science and Pseudoscience: In Defense of Demarcation Projects," which appears in *Science Unlimited* (Chicago: University of Chicago Press, 2017). Earlier salvos include physicist Richard Feynman's statement that "philosophy of science is as useful to scientists as ornithology is to birds" and physicist Steven Weinberg's entire chapter "Against Philosophy" in his book *Dreams of a Final Theory* (New York: Pantheon, 1992). One should balance all of this, however, against Einstein's quite high opinion of philosophy and its importance in scientific work. See Don A. Howard, "Albert Einstein as Philosopher of Science," *Physics Today* (Dec. 2005): 34–40.

6. Note that this is very different from the claim that a scientific theory is true. There is unfortunately no guarantee that a scientific theory is true, even if we are justified in believing it based on a rational consideration of the evidence. (We will discuss this issue further in chapter 2.)

7. Though note that in *Philosophy of Pseudoscience: Reconsidering the Demarcation Problem* (Chicago: University of Chicago Press, 2013), Massimo Pigliucci and Maarten Boudry self-consciously seek to resurrect the problem of demarcation. Also note that in *The Atheist's Guide to Reality: Enjoying Life without Illusions* (New York: Norton, 2012), Alex Rosenberg embraces the term "scientism" as a badge of honor.

8. Popper's model has also been handicapped by his occasional insistence that some scientific fields—like evolutionary biology—were not actually scientific, because they did not pass his criterion of demarcation. Though he later recanted this position, to many it seemed to reveal the hubris of thinking that there was a sharp theoretical line that could be drawn between science and nonscience. Popper made his claim that natural selection was "tautological" and "not a testable scientific theory," in his autobiography, which appears as part 1 of *The Philosophy of Karl Popper: The Library of Living Philosophers*, vol. 14, ed. Paul Schilpp (La Salle, IL: Open Court, 1974), 133–143. Within a few years Popper retracted this opinion, but he still held to the idea that it was "difficult to test" Darwinian theory. See his "Natural Selection and the Emergence of the Mind," *Dialectica* 32 (1978): 344.

9. It should be noted that Kuhn himself resisted this interpretation of his work. While he acknowledged the potential influence of such theoretical virtues as scope, simplicity, and fruitfulness on paradigm choice, he never gave up the idea that science was and should be evidence based (see quotation accompanying note 27 in chap. 3). For more on Kuhn's account of the role of subjective, factors in theory choice, see his "Objectivity, Value Judgment, and Theory Choice," in *The Essential Tension* (Chicago: University of Chicago Press, 1974), 320–339.

10. Imre Lakatos and Alan Musgrave, eds., *Criticism and the Growth of Knowledge* (Cambridge: Cambridge University Press, 1970); Paul Feyerabend, *Against Method* (London: Verso, 1978); Larry Laudan, *Progress and Its Problems: Towards a Theory of Scientific Growth* (Berkeley: University of California Press, 1978); Steve Fuller, *Philosophy of Science and Its Discontents* (New York: Guilford Press, 1992).

11. For further discussion of Popper's objections to a science of human behavior, see Lee McIntyre, *Laws and Explanation in the Social Sciences: Defending a Science of Human Behavior* (Boulder: Westview Press, 1996), 34–45, 64–75. Popper's own argument is spread over several of his works: *The Poverty of Historicism* (London: Routledge, 1957), *The Open Universe* (London: Routledge, 1982), and "Prediction and Prophecy in the Social Sciences," which appears in *Conjectures and Refutations: The Growth of Scientific Knowledge* (New York: Harper Torchbooks, 1965).

12. We will take up all these issues when we discuss denialism and pseudoscience in chapter 8.

1 Scientific Method and the Problem of Demarcation

1. There is some dispute over who first came up with the term *scientific method.* Many trace it back to the thirteenth-century philosopher and theologian Roger Bacon—not to be confused with the sixteenth-century philosopher Francis Bacon—who along with his teacher Robert Grosseteste first put forward the idea that scientific knowledge should be based on sensory evidence. Francis Bacon later advocated and refined this method in service of the same empirical goal.

2. See Noretta Koertge, ed., *New Dictionary of Scientific Biography* (New York: Scribner's, 2007).

3. Later in this chapter we will explore a few of the reasons that philosophers of science, such as Popper and Kuhn, have given for rejecting the idea of scientific method.

4. Laudan, "The Demise of the Demarcation Problem," in *Beyond Positivism and Relativism: Theory, Method, and Evidence,* ed. Larry Laudan (Boulder: Westview Press, 1996), 210–222. For an excellent general discussion, see also Thomas Nickles, "The Problem of Demarcation: History and Future," in *Philosophy of Pseudoscience,* ed. M. Pigliucci and M. Boudry (Chicago: University of Chicago Press, 2013), 101–120.

5. The classic source for Logical Positivism at its peak is A. J. Ayer's *Language, Truth, and Logic* (Mineola, NY: Dover, 1952). For an interesting history of its long slide afterward, see P. Achinstein and S. Barker, *The Legacy of Logical Positivism* (Baltimore: Johns Hopkins University Press, 1969), which goes into some depth about the particular problems the approach faced, especially when the Logical Positivists realized that some of their own statements could not pass a credible version of the verification test.

6. Chapter 2 will provide a longer discussion of the problem of induction and the issues it creates for scientific reasoning. In short, the problem of induction is that one cannot be certain of any statement that is open to later refutation by empirical evidence.

7. Strontium-90 is a radioactive particle produced by atomic fission that is absorbed into the environmental food cycle, where it competes with calcium and settles into the bones of those who ingest it. Before the 1963 nuclear test ban treaty, there were hundreds of atmospheric nuclear tests worldwide. In one study in St. Louis in 1963, baby teeth were found to have fifty times the level of Strontium-90 as those born in 1950. The half-life of Strontium-90 is 28 years.

8. Actually, it may not, for the problem of induction technically undermines not just the certainty but also the *probability* of inductive statements. For more discussion, see chapter 2.

9. Here we might still have to do some more investigation to figure out *why* Gabriel does not have Strontium-90 in his bones. Was he born before 1945? Was he born in 1994 but did not live near a nuclear reactor? But the point is that the reasoning is

deductively valid. Once we know that he does not have Strontium-90 in his bones we can rule out that he was born between 1945 and 1991.

10. Karl Popper, *Conjectures and Refutations* (New York: Harper Torchbooks, 1965), 36.

11. Karl Popper, *The Logic of Scientific Discovery* (New York: Basic Books, 1959). It is fascinating to note that although virtually every scholar characterizes Popper's concern in the demarcation debate as one between science and nonscience (or pseudoscience), Popper does not himself use either of these terms in this book. Where he comes closest to defining demarcation (in section 4), Popper says that it is meant to "distinguish between the empirical sciences on the one hand, and mathematics and logic as well as 'metaphysical' systems on the other." This makes clear that he did not intend science to be distinguished just from "metaphysical speculation" but also from other "unscientific" inquiries like math and logic, which is perhaps where the idea of nonscience comes from: it is an amalgam of pseudoscience and unscientific inquiries. Later in his *Conjectures and Refutations: The Growth of Scientific Knowledge* (New York: Harper Torchbooks, 1965), Popper begins to use the term "pseudo-science" more or less interchangeably with "metaphysics" as the sole contrast for science, with the term "nonscience" still unmentioned. So why use it now? First, it has come to be used as a category term in a debate that has outgrown Popper. Second, it seems faithful to Popper's original meaning. As we will see, however, *it makes a great deal of difference* whether we see the demarcation debate as one between science and nonscience, or between science and pseudoscience. I prefer the term "nonscience," even though it is not used explicitly by Popper.

12. The full text of Act 590 can be found in *But Is It Science? The Philosophical Question in the Creation/Evolution Controversy*, ed. M. Ruse (Amherst, NY: Prometheus Books, 1996), 283–286.

13. See Lee McIntyre, *Respecting Truth: Willful Ignorance in the Internet Age* (New York: Routledge, 2015), 64.

14. For more on the positive evidence in favor of evolution by natural selection, and the absence of such evidence for creation science and its successors, see pp. 40–41 and 175–180.

15. This was not, however, seen by all as an unfettered benefit. For doesn't it open the door that something like astrology might come up with a falsifiable prediction and therefore be accepted as science? Worse, to the extent that astrology has already been proven *false*, doesn't this show that it must be *falsifiable*, and so therefore scientific?

16. This essay appears as the preface to Popper's volume *Realism and the Aim of Science* (Lanham, MD: Rowman and Littlefield, 1983).

17. Popper, "Science: Conjectures and Refutations," in *Conjectures and Refutations*, 46.

18. Scientific method is just one way for there to be a methodological demarcation between science and nonscience. Falsification is another. Perhaps there are more.

As we have seen, there are also attempts to demarcate science from nonscience based on nonmethodological criteria, such as the Logical Positivist's idea to draw a distinction between statements that are cognitively meaningful and those that are cognitively meaningless (though since this uses the verification criterion, one could argue that this too was an attempt at methodological demarcation).

19. Kuhn had a hand in arguing against the simple idea of "scientific method," through his claim that all observation was theory laden. See his *The Structure of Scientific Revolutions* (Chicago: University of Chicago Press, 1962).

20. The closest Kuhn ever came to a criterion of demarcation was in his commentary on Popper entitled "Logic of Discovery or Psychology of Research," in *The Philosophy of Karl Popper*, vol. 14, ed. Paul Schilpp (La Salle, IL: Open Court, 1974), where Kuhn wrote "it is normal science, in which Sir Karl's sort of testing does not occur, rather than extraordinary science which most nearly distinguishes science from other enterprises. If a demarcation criterion exists (we must not, I think, seek a sharp or decisive one), it may lie just in that part of science which Sir Karl ignores" (802). But cf. here Tom Nickles, "Problem of Demarcation," 109, who calls Kuhn's idea of puzzle solving within normal science "Kuhn's criterion." See also Sven Ove Hansson's entry "Science and Pseudo-Science" in *The Stanford Encyclopedia of Philosophy*, in which he explicitly calls Kuhn's views on puzzle solving his "criterion of demarcation," https://plato.stanford.edu/entries/pseudo-science/.

21. Kuhn gives an account of "normal science"—which he argues is what the majority of scientists do most of the time—in *The Structure of Scientific Revolutions*.

22. In Feyerabend's book *Against Method* (London: Verso, 1975), he claims that "anything goes" in science and that it practices "methodological anarchy."

23. For an enjoyable history of these efforts, see Peter Achinstein and Stephen Barker, eds., *The Legacy of Logical Positivism: Studies in the Philosophy of Science* (Baltimore: Johns Hopkins University Press, 1969).

24. For instance, that it was possible to maintain strict objectivity and that there was an absolute distinction between facts and values.

25. Laudan, "Demise of the Demarcation Problem," 216–217.

26. All of this, of course, is a classic problem in any decision procedure, where we want to include all and only those things we favor and exclude all and only those we don't. Examples include deciding whether to shoot down a plane (is it enemy or ally?), allow an inference to be accepted (is it valid or invalid?), or remove a tumor (is it cancerous or benign?). A perfect decision procedure makes no errors (either false positives or false negatives), which is what many are searching for with a criterion of demarcation between science and nonscience. Unfortunately, the rate of false positives is inversely proportional to the rate of false negatives, and vice versa. By reducing one, we cannot help but increase the likelihood of the other.

27. Notice that Laudan is here making a sort of "meta-argument" not only that past philosophers have failed to solve the problem of demarcation, but that doing so would require providing the necessary and sufficient conditions for science. He is, in other words, making the idea of providing necessary and sufficient conditions its own *necessary condition* for solving the problem of demarcation. Presumably, he also believes that if one *could* provide the necessary and sufficient conditions of science, this would be sufficient for solving the problem of demarcation. Therefore, if one adds these two claims together, one gets the fascinating larger claim that one has solved the problem of demarcation if and only if one provides the necessary and sufficient conditions of science. Less artfully, this says that providing the necessary and sufficient conditions of science is itself a necessary and sufficient condition for solving the problem of demarcation. For more on the delicate question of necessary and sufficient conditions in the demarcation debate, see chapter 4.

28. Hansson, "Science and Pseudo-Science," *Stanford Encyclopedia of Philosophy*.

29. Laudan, "Demise of the Demarcation Problem," 218–219.

30. See Robert Feleppa, "Kuhn, Popper, and the Normative Problem of Demarcation," in *Philosophy of Science and the Occult*, ed. Patrick Grim (Albany: SUNY Press, 1990), 142. Popper writes, "a system is to be considered scientific only if it makes assertions which may clash with observations." *Conjectures and Refutations*, 256.

31. Recall here Popper's claim that evolutionary biology was not testable (see note 8 in the introduction).

32. This quotation is taken from Popper's "Falsifizierbarkeit, zwei Bedeutungen von" ([1989] 1994), 82, cited in Sven Hansson's article "Science and Pseudo-Science," in *The Stanford Encyclopedia of Philosophy*, https://plato.stanford.edu/entries/pseudo-science/.

33. Quotations are from Hannson, "Science and Pseudo Science." See also Frank Cioffi, "Psychoanalysis, Pseudoscience and Testability," in *Popper and the Human Sciences*, ed. Gregory Currie and Alan Musgrave (Dordrecht: Martinus Nijhoff, 1985), 13–44.

34. Larry Laudan, "Science at the Bar: Causes for Concern," in *Beyond Positivism and Relativism: Theory, Method, and Evidence* (Boulder: Westview Press, 1996), 223.

35. Tom Nickles, "Problem of Demarcation," 111.

36. McIntyre, *Respecting Truth*, 64–71.

37. McIntyre, *Respecting Truth*, 69. For further discussion, see chapter 8.

38. Massimo Pigliucci, "The Demarcation Problem: A (Belated) Response to Laudan," in *Philosophy of Pseudoscience*, 17–19.

39. Pigliucci, "Demarcation Problem," 22.

40. Pigliucci, "Demarcation Problem," 25.

41. Pigliucci, "Demarcation Problem," 25.

42. Sven Hansson, "Defining Science and Pseudoscience," in *Philosophy of Pseudoscience*, 61–77.

43. Maarten Boudry, "Loki's Wager and Laudan's Error," in *Philosophy of Pseudoscience*, 79–98.

44. Recall that although Popper does not use the term "nonscience," it is closest to his original meaning in *The Logic of Scientific Discovery*. See note 11 in this chapter.

45. Although Boudry does not us the term "unscience" in his essay, I am claiming that he ought to, for the term "nonscience" seems a misnomer for the territorial dispute he has in mind.

46. For a careful rendering of this distinction see Tom Nickles, "Problem of Demarcation," 101–120, and James Ladyman, "Toward a Demarcation of Science from Pseudoscience," 45–59, in *Philosophy of Pseudoscience*.

47. One might think of these taken together as constituting nonscience. See note 11 in this chapter.

48. See Laudan, "Demise of the Demarcation Problem."

2 Misconceptions about How Science Works

1. It is important to note that Popper took pains throughout his life to defend himself against the charge of naive falsificationism. For more, see chap. 2, note 10.

2. Popper, "Replies to My Critics," in *The Philosophy of Karl Popper*, vol. 14, ed. Paul Schilpp (La Salle, IL: Open Court, 1974), 984.

3. Some will be tempted to recall here the famous story that, upon learning that his theory had been confirmed, Einstein was asked what he would have done if the experiment had shown otherwise. He replied, "then I would have been sorry for the dear Lord; the theory is correct." Despite Einstein's confidence, does anyone really believe that if his prediction had been disproven the theory would have been accepted? Perhaps further tests would have been necessary, or it might have been revealed that Eddington's measurements were flawed, but if not the theory would at least had to have been modified.

4. Samir Okasha, *Philosophy of Science: A Very Short Introduction* (Oxford: Oxford University Press, 2016), 15.

5. This is the observation that in its revolution around the Sun, Mercury does not precisely retrace its orbit, but instead shifts by a minuscule amount each time. This could only be to the result of some gravitational force.

6. Kuhn, *The Structure of Scientific Revolutions* (Chicago: University of Chicago Press, 1962).

7. See Hilary Putnam, "The 'Corroboration' of Theories," in *The Philosophy of Karl Popper*, 223.

8. Popper uses this phrase many times in *The Logic of Scientific Discovery* (New York: Basic Books, 1959), including in a section heading called "The Positive Theory of Corroboration: How a Theory May 'Prove Its Mettle.'"

9. As Tom Nickles writes in "The Problem of Demarcation: History and Future," in *Philosophy of Pseudoscience*, ed. M. Pigliucci and M. Boudry (Chicago: University of Chicago Press, 2013): "In [Popper's] view even a general theory that has passed many severe tests (and is thus highly corroborated) still has probability zero. According to Popper…'in an infinite universe…*the probability of any (non-tautological) universal law will be zero.*' (Popper's emphasis)" (107–108).

10. "Falsifiability as a criterion of demarcation is a *purely logical affair* and does not depend on our (non-existent) empirical or practical ability to falsify a statement conclusively" (emphasis in original). Personal communication from Karl Popper to Lee McIntyre, March 26, 1984.

11. See Popper, *Conjectures and Refutation* (New York: Harper Torchbooks, 1965), 41, note 8.

12. For instance, the notions of phlogiston, ether, and caloric.

13. Richard Feynman, "The Essence of Science in 60 Seconds," https://www.youtube.com/watch?v=5v8habYTfHU.

14. This is to say that even if falsification fails as a criterion of demarcation, it may get something right about science, by focusing our attention on the idea that comparing a theory with evidence is crucial to what is special about science. Remember too that, despite those who would seem to deliberately misunderstand him, Kuhn also champions the role of evidence in science.

15. Another example, which we will explore in chapter 3, is Semmelweis's discovery of the cause of childbed fever, which was later supported by the germ theory of disease.

16. It is an interesting question, though, whether what defeated Bode's law was its lack of theory or its later failed predictions. See McIntyre, "Accommodation, Prediction, and Confirmation," *Perspectives on Science* 9, no. 3 (2001): 308–323.

17. See Alberto Guijosa, "What Is String Theory?" http://www.nuclecu.unam.mx/~alberto/physics/string.html.

18. The most readable book on this subject is Brian Greene's *The Elegant Universe: Superstrings, Hidden Dimensions, and the Quest for the Ultimate Theory* (New York: Norton, 2010).

19. It is interesting to note here that "string theory" was originally called the "string hypothesis." See Ethan Siegel, "Why String Theory Is Not a Scientific Theory," Forbes .com, Dec. 23, 2015, http://www.forbes.com/sites/startswithabang/2015/12/23/why -string-theory-is-not-science/.

20. Richard Dawid, *String Theory and Scientific Method* (Cambridge: Cambridge University Press, 2014). Some of this may sound like Kuhn, but it is important to note that Kuhn's "extra-empirical" criteria were intended to supplement empirical evidence, not substitute for it.

21. See in particular remarks at the conference by David Gross (a Nobel Prize–winning physicist) who "classified string theory as testable 'in principle' and thus perfectly scientific, because the strings are potentially detectable." David Castelvecchi, "Is String Theory Science?" *Nature*, Dec. 23, 2015, https://www.scientificamerican.com /article/is-string-theory-science/. For more perspective on the conference see Natalie Wolchover, "Physicists and Philosophers Hold Peace Talks, If Only for Three Days," *Atlantic*, Dec. 22, 2015, https://www.theatlantic.com/science/archive/2015/12/physics -philosophy-string-theory/421569/.

22. See Lee Smolin, *The Trouble with Physics: The Rise of String Theory, the Fall of a Science, and What Comes Next* (New York: Mariner, 2007), and Peter Woit, *Not Even Wrong: The Failure of String Theory and the Search for Unity in Physical Law* (New York: Basic, 2007).

23. Peter Woit, "Is String Theory Even Wrong?" *American Scientist* (March–April 2002), http://www.americanscientist.org/issues/pub/is-string-theory-even-wrong/.

24. Theodosius Dobzhansky, "Nothing in Biology Makes Sense except in Light of Evolution," *American Biology Teacher* 35, no. 3 (March 1973): 125–129.

25. See also Larry Laudan, *Progress and Its Problems: Towards a Theory of Scientific Growth* (Berkeley: University of California Press, 1978).

26. For an account of how to explain the eye, and other "organs of extreme perfection," see Richard Dawkins, *Climbing Mount Improbable* (New York: Norton, 2016). For an account of the relatively recent discovery of a fossilized fish with shoulders, elbows, legs, a neck, and wrists—named Tiktaalik—see Joe Palca, "The Human Edge: Finding Our Inner Fish," NPR, http://www.npr.org/2010/07/05/127937070/the-human-edge -finding-our-inner-fish.

27. See the discussion of intelligent design in chapter 8.

28. There is another virtue for this point of view as well, which is that it allows us to dispute the idea that scientific theories are weak unless they can be proven. While it is true that you cannot prove an empirical statement, this does not mean that we are justified in believing whatever we want because it "might" be true. The Flat Earth hypothesis "might" be true, but where is the evidence? Even if one guessed correctly (like the proverbial stopped clock that is right twice a day), this is not science. To say

that we are not warranted in believing "anything" we want does not mean that we are warranted in believing "nothing." Science is about having a theory that is tested against the evidence, which is what gives us justification for believing in it.

29. This is usually called *pessimistic induction*. Note, though, that this should be carefully distinguished from something called *counterinduction*, which is the idea that because something has happened in the past, it is bound to fail in the future. Consider the inductivist dice player who has rolled three sevens in a row and says "I'm bound to do it again!" This is not valid, but neither is his counterinductivist friend who says "no you've used up all your luck; it's nearly impossible you'd do it again." If the dice are fair, the odds are the same for each throw, no matter what has happened in the past. By contrast, pessimistic induction is merely the observation that, given the way induction works, it is nearly inconceivable to think that our limited experience will lead us to discover a true theory; thus, virtually every scientific theory will someday be overthrown.

30. One of the most intriguing essays on this subject was written by the eminent pragmatist philosopher Charles S. Peirce. In "The Scientific Attitude and Fallibilism," Peirce links the sorts of "virtuous characteristics" we might associate with someone who has the scientific attitude with respect for the idea that empirical knowledge must always be incomplete. The person who has the "scientific attitude" must never block the way of inquiry by thinking that their knowledge is complete, because in a field as open-ended as science there is always more to learn. See https://www.textlog.de/4232.html.

31. See D. H. Mellor, "The Warrant of Induction," in *Matters of Metaphysics* (Cambridge: Cambridge University Press, 1991), https://www.repository.cam.ac.uk/bitstream/handle/1810/3475/InauguralText.html?sequence=5.

32. One might wish, however, to allow that it is perfectly acceptable for scientists to continue to use the *word* "true"—without having always to insert the fallibilist asterisk that future evidence may show their belief to be false—as long as one remembers that in every discussion of empirical knowledge this is surely the case.

33. David Hume, *A Treatise of Human Nature* (London, 1738), Book VII.

34. Although Reichenbach's own work is enlightening, the most approachable discussion of this idea can be found in Wesley Salmon, "Hans Reichenbach's Vindication of Induction," *Erkenntnis* 35, no. 1 (July 1991): 99–122.

35. It is crucial to note that the word "vindicated" is used here, as opposed to saying that induction is "verified."

36. The principle of conservativism in changing one's theory is a deeply held norm in science. As we have seen, Popper held that—other things being equal—we ought to give heavy deference to those theories that have survived longer or are better "entrenched." Quine too has held (on practical grounds) that we may legitimately prefer those ideas that do least violence to our other existing beliefs.

37. There are still several problems to be worked out here. For one thing, even if pragmatic vindication may provide a bulwark against certainty, what can it say about whether warranted beliefs are more *probable* in light of the evidence? Remember that probability is undermined by the inductive observation that we cannot be sure that the piece of the world we have sampled so far is representative of the rest of it. But couldn't we say that, if it is not, then no belief is warranted, but if it is then our belief is at least as well warranted as any other? For further details, see my "A Pragmatic Vindication of Warranted Belief" (work in progress).

38. Descartes worried that our senses were potentially unreliable owing to the possibility that we were perpetually dreaming or could be duped by an evil genius. Goodman—through his famous "New Riddle of Induction"—worried that we could not even be sure that the predicates we were using (like "green" and "blue") were superior to fabricated ones (like "grue" and "bleen") that may be equally well-supported by empirical evidence. See Nelson Goodman, *Fact, Fiction, and Forecast* (Cambridge, MA: Harvard University Press, 1955).

39. If the point of defending the warrant of induction is not just to show that scientists know X, but to show that they must *know that they know* X, isn't this just to give in to the demand for certainty that is based on one of the misconceptions about science?

40. See, e.g., the letter signed by 255 members of the US National Academy of Sciences that I mentioned on the first page of this book.

3 The Importance of the Scientific Attitude

1. For the classic argument, see Norwood R. Hanson, *Patterns of Discovery: An Inquiry into the Conceptual Foundations of Science* (Cambridge: Cambridge University Press, 1958). For an excellent overview, see Thomas Nickles, "Introductory Essay: Scientific Discovery and the Future of Philosophy of Science," in *Scientific Discovery, Logic, and Rationality*, ed. T. Nickles (Dordrecht: Reidel, 1980), 1–59.

2. The quotation above is widely attributed to Feynman, but it is only a piece of what he had to say on the subject. For more see his delightful essay "What Is Science?" *Physics Teacher* 7, no. 6 (1968): 313–320.

3. Some would maintain that the problem here is much more severe—that the evidence is *always* ambiguous because there are infinitely many possible theories that could in principle fit the same data. See Helen Longino, "Underdetermination: A Dirty Little Secret?" *STS Occasional Papers 4* (London: Department of Science and Technology Studies, University College London, 2016). For more background, see Paul Horwich, "How to Choose between Empirically Indistinguishable Theories," *Journal of Philosophy* 79, no. 2, (1982): 61–77, and Larry Laudan and Jarrett Leplin, "Empirical Equivalence and Underdetermination," *Journal of Philosophy* 88, no. 9 (1991): 449–472. For a general discussion of this problem, see Lee McIntyre, "Taking Underdetermination Seriously," *SATS: Nordic Journal of Philosophy* 4, no. 1 (2003): 59–72.

4. I will have more to say about the practices by which scientists do this in chapter 5.

5. Though we can certainly look for some telltale markers of poor reasoning. Denialists and pseudoscientists may *say* that they care about evidence, but it matters a great deal how one demonstrates this. Wishful thinkers may *hope* that there is evidence to back up their theory, but this is not sufficient. Those who engage in hasty generalizations are likewise disrespecting what it means to care about evidence. To give lip service to empirical support—for instance by cherry picking only the data that fit one's theory—is not to truly care about evidence. One demonstrates care for scientific evidence by working hard to see not just whether there is evidence that fits a theory, but also whether there is anything that is capable of refuting it. For more on this issue, see the discussion of denialism and pseudoscience in chapter 8.

6. And occasionally a whole scientific field can get off track. What does one do when the individual is right and the scientific community is wrong? This issue will be discussed in chapter 8 with the example of Harlan Bretz.

7. Peter Achinstein, ed., *Scientific Evidence: Philosophical Theories and Applications* (Baltimore: Johns Hopkins University Press, 2005), 1. Some of the different concepts of evidence that Achinstein discusses include (1) falsificationist, (2) inductivist, (3) explanationist, (4) Bayseian, and (5) "anarchist" accounts.

8. See in particular Deborah Mayo, *Error and the Growth of Experimental Knowledge* (Chicago: University of Chicago Press, 1996). Mayo's "error-statistical" model offers a stirring challenge to the dominant Bayesian approach. For other perspectives, see Peter Achinstein, *The Book of Evidence* (Oxford: Oxford University Press, 2003), and Clark Glymour, *Theory and Evidence* (Princeton: Princeton University Press, 1980).

9. Roughly speaking, adherents of subjectivist approaches feel that it is reasonable to rely on our background knowledge of prior probabilities as a starting point for assessing the likelihood of a hypothesis, with modification over time based on experience. Frequentists dispute this and argue that it is folly to attach a probability estimate to a hypothesis before it has been tested against experience. Mayo argues that scientific knowledge arises only from "severe tests" of a theory.

10. I will have much more to say about this issue in chapter 4.

11. One such example occurred recently when astronomers argued that the BICEP2 telescope at the South Pole had discovered "direct evidence that in the first fraction of a second after the big bang, the universe underwent a bizarre exponential growth spurt." It was later shown that the finding was an artifact of microwave radiation from background dust. At first, the researchers were reluctant to abandon their conclusion, but later came around when the evidence became overwhelming. Adrian Cho, "Curtain Falls on Controversial Big Bang Result," *Science*, Jan. 30, 2015, http://www.sciencemag.org/news/2015/01/curtain-falls-controversial-big-bang-result.

12. I will have much more to say about the critical role of community judgment in the evaluation of scientific theories in chapter 5.

13. Note that these brief examples are here included to illustrate what I mean by the scientific attitude. Much more detail will be provided in chapter 6 (where I will discuss how the scientific attitude transformed modern medicine) and in chapter 5 (where I will discuss how cold fusion undermined the scientific attitude by refusing to facilitate group scrutiny of individual work).

14. Carl Hempel, *Philosophy of Natural Science* (New York: Prentice Hall, 1966), 3–8.

15. See Noretta Koertge's fascinating essay, "Belief Buddies versus Critical Communities," in *Philosophy of Pseudoscience*, ed. M. Pigliucci and M. Boudry (Chicago: University of Chicago Press, 2013), 165–180, where she argues that the critical appraisal of one's ideas by one's scientific peers is valuable in improving them and bringing them to the attention of the wider community.

16. Roy Porter, *The Greatest Benefit to Mankind: A Medical History of Humanity* (New York: Norton, 1999), 369.

17. Porter, *Greatest Benefit*, 369.

18. Porter, *Greatest Benefit*, 369–370; W. F. Bynum et al., *The Western Medical Tradition 1800–2000* (Cambridge: Cambridge University Press, 2006), 156; Hempel, *Philosophy of Natural Science*, 3–8.

19. Another example of a lone scientist who was vindicated after years of resistance is Galileo. For a more modern example—which bears on the question of what happens to the scientific attitude when the individual is right but the group is wrong—see my discussion in chapter 8 of Harlen Bretz and his theory of megafloods in the Eastern Washington scablands.

20. Koertge, "Belief Buddies," raises the intriguing point, though, that even the lone genius's ideas could be improved by membership in a critical community.

21. For further discussion of cold fusion within the context of peer review, see chapter 5. See also Lee McIntyre, *Dark Ages: The Case for a Science of Human Behavior* (Cambridge, MA: MIT Press, 2006), 19–20.

22. See also Robert Merton's important essay "The Normative Structure of Science" (1942) (reprinted as chapter 13 in *The Sociology of Science*, ed. Robert Merton [Chicago: University of Chicago Press, 1973]), which has some enormously insightful things to say about the role and importance of values in scientific inquiry.

23. This idea may be more important than it sounds. Is it possible that Popper missed the point of his own criterion of demarcation? Perhaps what matters most is not whether a theory is falsifiable, but whether the scientists who put it forward

seek to falsify it. For an intriguing take on this idea, see Janet Stemwedel's essay "Drawing the Line between Science and Pseudo-Science," where she writes, "The big difference Popper identifies between science and pseudo-science is a difference in attitude. While a pseudo-science is set up to look for evidence that supports its claims…a science is set up to *challenge* its claims and look for evidence that might prove it false. In other words, pseudo-science seeks confirmations and science seeks falsifications." *Scientific American*, Oct. 4. 2011, https://blogs.scientificamerican.com/doing-good-science/drawing-the-line-between-science-and-pseudo-science/.

24. Karl Popper, "Remarks on the Problems of Demarcation and of Rationality," in *Problems in the Philosophy of Science*, ed. Imre Lakatos and Alan Musgrave (Amsterdam: North-Holland, 1968), 94.

25. *The Philosophy of Karl Popper*, ed. P. A. Schilpp (LaSalle: Open Court, 1974), 29.

26. K. Brad Wray, "Kuhn's Social Epistemology and the Sociology of Science," in *Kuhn's Structure of Scientific Revolutions—50 Years On*, ed. W. Devlin and A. Bokulich (Dordrecht: Springer, 2015), 175–176.

27. Thomas Kuhn, *The Road since Structure: Philosophical Essays, 1970–1993, with an Autobiographical Interview*, ed. J. Conant and J. Haugeland (Chicago: University of Chicago Press, 2002), 101.

28. Though some arguably did. Note again Merton's discussion of the values of science. He lists four of them (Communalism, Universalism, Disinterestedness, and Organized Skepticism), which Sven Hansson—in his essay "Science and Pseudo Science" in the *Stanford Encyclopedia of Philosophy*—says have been underappreciated for their role in the demarcation debate. It is intriguing to note that Kuhn says some very laudatory things about Merton in his *Structure of Scientific Revolutions*.

29. Note here Kuhn's statement (see page 211, note 20) where he says that we must not try too hard to look for a criterion of demarcation.

30. Popper, "Science: Conjectures and Refutations," in *Conjectures and Refutations* (New York: Harper Torchbooks, 1965), 52.

31. See http://www.earlymoderntexts.com/assets/pdfs/bacon1620.pdf.

32. Rose-Mary Sargent, "Virtues and the Scientific Revolution," in *Scientific Values and Civic Virtues*, ed. Noretta Koertge (Oxford: Oxford University Press, 2005), 78.

33. Noretta Koertge, ed., *Scientific Values and Civic Virtues* (Oxford: Oxford University Press, 2005), 10.

34. Alasdair MacIntyre, *After Virtue: A Study in Moral Theory* (South Bend: University of Notre Dame Press, 1981), 1.

35. In their important paper "The Virtues of Scientific Practice: MacIntyre, Virtue Ethics, and the Historiography of Science," Daniel Hicks and Thomas Stapleford

recognize that "although history and philosophy of science is not the focus of MacIntyre's virtue ethics, science serves as a prominent exemplar in his writing." *Isis* 107, no. 3 (Sept. 2016): 4.

36. As we will see in chapter 8, it is also possible for the scientific community to make a mistake.

37. At the end of their paper, Hicks and Stapleford argue for precisely this sort of vision, where the virtues of science are recognized for their role in community practice, by analogy with MacIntyre's vision for virtue ethics.

38. Abrol Fairweather, ed., *Virtue Epistemology Naturalized: Bridges between Virtue Epistemology and Philosophy of Science* (Dordrecht: Springer, 2014).

39. Is there a relationship between the content of a theory and the behavior of the people who advance it? Traditionally the problem of demarcation has focused on the former, yet perhaps there is some role for the latter. See here Martin Curd's review of Pigliucci and Boudry's *Philosophy of Pseudoscience* in *Notre Dame Philosophical Reviews* (July 22, 2014), where, citing Boudry's article, Curd raises the question of whether the *behavior* of pseudo-scientists is relevant to the question of demarcation: http://ndpr.nd.edu/news/philosophy-of-pseudoscience-reconsidering-the-demarcation-problem/. Action, after all, is motivated by values. How one approaches evidence matters. One's intentions matter. This seems consonant with the scientific attitude.

40. We will explore these practices of science in detail in chapter 5.

41. Clearly, we must also care about the logical coherence and consistency of our theory. If a theory is self-contradictory, no evidence can save it.

4 The Scientific Attitude Need Not Solve the Problem of Demarcation

1. See my earlier discussion of Laudan's "meta-argument" in chapter 1, note 27, above.

2. In logic, "if A then B" added to "if B then A" gets you "A if and only if B," which is to assert the logical equivalence of A and B. If the reader wants to learn more about the intricacies of how logicians use terms like "necessary and sufficient conditions"; the relationships between "if," "only if," and "if and only if"; and terms like "biconditional," "contraposition," and "logical equivalence," there are a host of textbooks in introductory logic, none better than E. J. Lemmon, *Beginning Logic* (Cambridge, MA: Hackett, 1978).

3. See discussion in chapter 1. Feleppa argues it was only intended to be a necessary criterion ("Kuhn, Popper, and the Normative Problem of Demarcation," in *Philosophy of Science and the Occult*, ed. Patrick Grim [Albany: SUNY Press, 1990]), but note Popper's (later) statement that it was a necessary and sufficient criterion.

4. Recall here Popper's claim that evolutionary biology was not falsifiable.

5. Popper, "Falsifizierbarkeit, zwei Bedeutungen von" ([1989], 1994), 82.

6. Laudan seems to think that the reason for offering both necessary and sufficient conditions is that each is needed to protect from criticisms that could be leveled against the other criterion alone. But maybe that is not the case, and instead, it just opens one's criterion up to attack from both sides.

7. Which is to say that they are true under identical circumstances. If something satisfies the criterion of being science, then it satisfies the criterion of being falsifiable, and vice versa.

8. Here the point just made against Laudan in note 6 of this chapter seems relevant. How does offering the sufficiency standard for falsifiability protect Popper from the criticism that his necessity criterion rules that "evolution biology is not science?" Likewise, how does the necessity standard for falsifiability protect him against criticism that his sufficiency criterion allows that astrology as a science?

9. But just imagine the difficulty of finding a criterion to measure *this*!

10. See Karl Popper, *The Logic of Scientific Discovery* (New York: Basic Books, 1959); Karl Popper, *Conjectures and Refutations* (New York: Harper Torchbooks, 1965); Larry Laudan, "The Demise of the Demarcation Problem," in *Beyond Positivism and Relativism* (Boulder: Westview Press, 1996).

11. See Boudry, "Loki's Wager and Laudan's Error," 80–82, and Hansson, "Defining Pseudoscience and Science," 61–77, both in *The Philosophy of Pseudoscience: Reconsid ering the Demarcation Problem*, ed. M. Pigliucci and M. Boudry (Chicago: University of Chicago Press, 2013). Is Pigliucci guilty of this as well? In his essay "The Demarcation Problem," he gives up on solving the problem of demarcation via necessary and sufficient conditions, and says that the whole approach of trying to come up with a criterion of demarcation is "fuzzy," arguably because he is concerned with demarcating between science and pseudoscience rather than science and nonscience.

12. See figure 1.1. For a thorough discussion of this problem, see the discussion in chapter 1. See also Tom Nickles, "The Problem of Demarcation: History and Future," 101–120, and James Ladyman, "Toward a Demarcation of Science from Pseudoscience," 45–49, both in *Philosophy of Pseudoscience*, ed. M. Pigliucci and M. Boudry (Chicago: University of Chicago Press, 2013).

13. Galileo famously said, "the intention of the Holy Ghost is to teach us how one goes to heaven, not how heaven goes." *A Letter to the Grand Duchess Christina of Tuscany* (Florence, 1615).

14. I suppose one might here try to draw a firmer line *within* the category of nonscience, between what I have called "unscience" and "pseudoscience." Although this would not amount to a larger criterion of demarcation between science and nonscience (or between science and pseudoscience), perhaps it would be useful to say

that within the category of nonscience there are those fields that do not *purport* to care about empirical evidence (literature, art), and those that *do* (astrology, creationism)...even if one could then argue that, as a matter of practice, the latter do *not* actually care. Even this modest step, however, might prove problematic, for it raises the temptation to (1) try to use this as leverage to change the subject in the larger demarcation debate or (2) go down the rabbit hole of looking for all of the necessary and sufficient conditions for some *new* demarcation debate within the category of nonscience. But if we have already had trouble with how "caring about evidence" might serve as a sufficient condition for demarcating science from nonscience, how difficult might it be to build another account based on *purporting* to care about evidence? Again, I think that the proper focus for the demarcation debate is between science and nonscience.

15. And, arguably, in evolutionary biology. See Michael Ruse, "Evolution: From Pseudoscience to Popular Science, from Popular Science to Professional Science," in *Philosophy of Pseudoscience*, 225–244.

16. See the excellent treatment of this subject in Frank Cioffi's essay "Pseudoscience: The Case of Freud's Sexual Etiology of the Neuroses," in *Philosophy of Pseudoscience*, 321–340.

17. I am indebted to Rik Peels for this example.

18. Although a case could be made that my brother has the scientific attitude, most philosophers would still want to say that he was not doing science. Sidney Morgenbesser once addressed this by saying that there may be a difference between something's being scientific and its being science. Boudry also raises this pragmatic issue in his essay "Plus Ultra: Why Science Does Not Have Limits," *Science Unlimited* (Chicago: University of Chicago Press, 2017).

19. Is plumbing a science? Boudry (in his essay "Plus Ultra") says it is all right to believe this—that there need be no epistemological limit to science. In his own essay "In Defense of Demarcation Projects," Pigliucci disagrees, but his reason for this is that scientists have a "fairly clearly defined role." Here I wholeheartedly agree with Boudry. What seems important is the approach that one takes in seeking "everyday knowledge," not some sociological fact about plumbers.

20. The only way for a demarcationist to avoid the "looking for one's keys" problem is either to refuse to embrace a sufficiency standard in the first place (in which case they might fail to be a demarcationist) or come up with more necessary criteria. But once we start down this road there seems no end to it, which is another reason why I prefer to avoid offering the scientific attitude as a sufficiency standard.

21. It is of course also missing from those disciplines like math and logic, which do not care about empirical evidence.

22. Emile Durkheim, *The Rules of Sociological Method* (Paris, 1895; author's preface to the second edition).

23. One of the most intriguing proposals I have read recently is made by Tom Nickles ("The Problem of Demarcation," in *Philosophy of Pseudoscience,* 116–117) who critically examines "fertility" as a possible criterion for demarcation. Maybe what is wrong with creationism, for instance, is not that it is false, pretending, or doesn't have the scientific attitude, but that it is just not very interested in giving us future scientific puzzles to solve.

24. Perhaps Popper's original instinct was right and the necessity condition is all one needs for a criterion of demarcation.

25. Hansson, "Defining Pseudoscience and Science," in *Philosophy of Pseudoscience,* 61.

26. See again my discussion of Laudan's meta-argument (chapter 1, note 27) in which he claims that in order to solve the problem of demarcation one would have to find the necessary and sufficient conditions for science.

27. See Pigliucci, "The Demarcation Problem," 21. But does he do this because he is once again equivocating between whether the target should be pseudoscience or nonscience?

28. This appears in *Science Unlimited,* ed. M. Boudry and M. Pigliucci (Chicago: University of Chicago Press, 2017). In this volume, Boudry and Pigliucci say that they are now concerned with a different demarcation dispute, this time more along the lines of what Boudry earlier called the territorial problem. In their earlier book, Pigliucci and Boudry were concerned with keeping pseudoscience from infecting science. Now they seem concerned with the problem of scientism: whether other areas of inquiry need to be protected *from* science.

29. Pigliucci, *Science Unlimited,* 197.

30. Kitcher says much the same thing about pseudoscience. "Pseudoscience is just what [pseudoscientists] do." Quoted in Boudry, "Loki's Wager," 91.

31. McIntyre, *Respecting Truth: Willful Ignorance in the Internet Age* (New York: Routledge, 2015), 107–109.

32. Pigliucci acknowledges as much in his essay "The Demarcation Problem," where he seems to agree with Boudry that the appropriate task for demarcation is to limn the difference between science and *pseudo*science, not science and *non*science.

33. See here the discussion in chapter 8 on the question of whether it is the content of a theory, or the behavior of the people who advance it, that makes the difference in science.

34. As we have seen with the examples of Galileo and Semmelweis, the scientific community is sometimes woefully irrational.

35. Pigliucci, *Science Unlimited,* 197.

36. For more on the interesting question of sorting out the family relations of different areas of inquiry when compared to science, see Tom Nickles, "Problem of Demarcation," and James Ladyman, "Toward a Demarcation of Science from Pseudoscience," 45–59, in *Philosophy of Pseudoscience.*

5 Practical Ways in Which Scientists Embrace the Scientific Attitude

1. James Ladyman, "Toward a Demarcation of Science from Pseudoscience," in *The Philosophy of Pseudoscience: Reconsidering the Demarcation Problem*, ed. Massimo Pigliucci and Maarten Boudry (Chicago: University of Chicago Press, 2013), 56.

2. See Noretta Koertge, "Belief Buddies versus Critical Communities," in *The Philosophy of Pseudoscience: Reconsidering the Demarcation Problem*, ed. Massimo Pigliucci and Maarten Boudry, 165–180 (Chicago: University of Chicago Press, 2013).

3. When it does, there is even a mechanism by which it can be retracted. See chapter 5 for more discussion of the issue of retraction in science.

4. See chapter 5 for more on p-hacking.

5. Robert Trivers, *The Folly of Fools: The Logic of Deceit and Self-Deception in Human Life* (New York: Basic Books, 2011).

6. Robert Trivers, "Fraud, Disclosure, and Degrees of Freedom in Science," *Psychology Today* (blog entry: May 10, 2012), https://www.psychologytoday.com/blog/the-folly-fools/201205/fraud-disclosure-and-degrees-freedom-in-science.

7. J. Wicherts et al., "Willingness to Share Research Data Is Related to the Strength of the Evidence and the Quality of Reporting of Statistical Results," *PLOS ONE* 6, no. 11 (Nov. 2011): e26828, http://journals.plos.org/plosone/article?id=10.1371/journal.pone.0026828.

8. J. Simmons et al., "False-Positive Psychology: Undisclosed Flexibility in Data Collection and Analysis Allows Presenting Anything as Significant," *Psychological Science* 22 (2011): 1359–1366, http://journals.sagepub.com/doi/pdf/10.1177/0956797611417632.

9. Simmons et al., "False-Positive Psychology," 1359.

10. Simmons et al., "False-Positive Psychology," 1360.

11. Daniel Kahneman, *Thinking Fast and Slow* (New York: Farrar, Straus and Giroux, 2013). Another modern classic in the field of behavioral economics is Richard Thaler, *Nudge: Improving Decisions about Health, Wealth, and Happiness* (New York: Penguin, 2009).

12. One good example of this is the cold fusion debacle, which we will explore in depth later in this chapter.

13. P. Wason, "Reasoning about a Rule," *Quarterly Journal of Experimental Psychology* 20, no. 3 (1968): 273–281.

14. Lee McIntyre, *Respecting Truth: Willful Ignorance in the Internet Age* (New York: Routledge, 2015), 15–16.

15. Cass Sunstein, *Infotopia: How Many Minds Produce Knowledge* (Oxford: Oxford University Press, 2008).

16. For a full discussion, see McIntyre, *Respecting Truth*, 117–118.

17. See the discussion in Sunstein, *Infotopia*, 207; McIntyre, *Respecting Truth*, 119.

18. Tom Settle, "The Rationality of Science versus the Rationality of Magic," *Philosophy of the Social Sciences* 1 (1971): 173–194, http://journals.sagepub.com/doi/pdf/10.1177/004839317100100201.

19. Settle, "The Rationality of Science," 174.

20. Sven Hansson, "Science and Pseudo-Science," *The Stanford Encyclopedia of Philosophy*.

21. Settle, "The Rationality of Science," 183.

22. Koertge, "Belief Buddies," 177–179.

23. Helen Longino, *Science as Social Knowledge: Values and Objectivity in Scientific Inquiry* (Princeton: Princeton University Press, 1990).

24. Longino, *Science as Social Knowledge*, 66–67.

25. Longino, *Science as Social Knowledge*, 69, 74.

26. Longino, *Science as Social Knowledge*, 216. For a fascinating contrast, one might consider Miriam Solomon's *Social Empiricism* (Cambridge, MA: MIT Press, 2001), in which she embraces the idea that the actions of the "aggregate community of scientists" trump how "individual scientists reason" (135), but disagrees with Longino on the question of whether such social interactions can "correct" individual biases in reasoning (139). While Solomon finds much to praise in Longino's point of view, she criticizes her for not embedding her "ideal" claims in more actual scientific cases. Indeed, Solomon maintains that when one looks at scientific cases it becomes clear that biases—even cognitive biases—play a *positive* role in facilitating scientific work. Far from thinking that science should eschew bias, Solomon makes the incredible claim that "scientists typically achieve their goals through the aid of 'biased' reasoning" (139).

27. Some of these errors, of course, are the result of cognitive biases that all human beings share. The point here is not that scientists do not have these biases, but that science writ large is committed to reducing them through group scrutiny of

individual ideas. For more on this, see Kevin deLaplante's Critical Thinker Academy podcast: https://www.youtube.com/watch?v=hZkkY2XVzdw&index=5&list=PLCD69C3C29B645CBC.

28. As noted, in chapter 7 we will discuss at length the issue of scientific misconduct that amounts to outright deception, such as data fabrication and manipulation.

29. G. King, R. Keohane, and S. Verba, *Designing Social Inquiry: Scientific Inference in Qualitative Research* (Princeton: Princeton University Press, 1994). Where possible, it is helpful to use statistical evidence in science. Where this is not possible, however, this is no excuse for being less than rigorous in one's methodology.

30. As expected, significance effects in published research tend to cluster at 5 percent.

31. For a comprehensive and rigorous look at a number of foundational issues in statistics and its relationship to the philosophy of science, one could do no better than Deborah Mayo's classic *Error and the Growth of Experimental Knowledge* (Chicago: University of Chicago Press, 1996). Here Mayo offers not only a rigorous philosophical examination of what it means to learn from evidence, but also details her own "error-statistical" approach as an alternative to the popular Bayesian method. The essential question for scientific reasoning is not just *that* one learns from empirical evidence but *how*. Mayo's championing of the role of experiment, severe testing, and the search for error is essential reading for those who wish to learn more about how to defend the claims of science as they pertain to statistical reasoning.

32. It is important to remember, though, that there is still a nonzero chance that even the most highly correlated events are causally unrelated.

33. The term *p-hacking* was coined by Simmons et al. in their paper "False Positive Psychology."

34. M. Head, "The Extent and Consequences of P-Hacking in Science," *PLOS Biology* 13, no. 3 (2015): e1002106, http://journals.plos.org/plosbiology/article?id=10.1371/journal.pbio.1002106.

35. One used to have to compute t-tests and f-tests by hand, then look up the corresponding p-value in a table.

36. Simmons et al., "False Positive Psychology," 1359.

37. Simmons et al., "False Positive Psychology," 1359.

38. Steven Novella, "Publishing False Positives," *Neurologica* (blog), Jan. 5, 2012, http://theness.com/neurologicablog/index.php/publishing-false-positives/.

39. Christie Aschwanden, "Science Isn't Broken: It's Just a Hell of a Lot Harder Than We Give it Credit For," *FiveThirtyEight*, Aug. 19, 2015, https://fivethirtyeight.com/features/science-isnt-broken/.

40. J. Ioannidis, "Why Most Published Research Findings Are False," *PLOS Medicine* 2, no. 8 (2005): e124, http://robotics.cs.tamu.edu/RSS2015NegativeResults/pmed.0020124.pdf.

41. See Ronald Giere, *Understanding Scientific Reasoning* (New York: Holt, Rinehart, and Winston, 1984), 153.

42. R. Nuzzo, "Scientific Method: Statistical Errors," *Nature* 506 (2014): 150–152.

43. Head, "Extent and Consequences of P-Hacking."

44. Nuzzo, "Scientific Method."

45. Nuzzo, "Scientific Method."

46. "Scientists' overreliance on p-values has led at least our journal to decide it has had enough of them. In February [2015], *Basic and Applied Psychology* announced that it will no longer publish p-values. 'We believe that the p [is less than] .05 bar is too easy to pass and sometimes serves as an excuse for lower quality research,' the editors wrote in their announcement. Instead of p-values, the journal will require 'strong descriptive statistics, including effect sizes.'" Aschwanden, "Science Isn't Broken."

47. Head, "Extent and Consequences." One should note, though, that Head's findings have been disputed by some who claim that they are an artifact of rounding at the second decimal. This isn't to say that p-hacking doesn't occur, just that the bump that Head purports to find at 0.05 in her p-curves is not good evidence that it is quite so widespread. C. Hartgerink, "Reanalyzing Head et al. (2015): No Widespread P-Hacking After All?" *Authorea*, Sept. 12, 2016, https://www.authorea.com/users/2013/articles/31568/_show_article.

48. S. Novella, "P-Hacking and Other Statistical Sins," *Neurologica* (blog), Feb. 13, 2014, http://theness.com/neurologicablog/index.php/p-hacking-and-other-statistical-sins/.

49. Or, in some cases, how it is measured. If one were a Bayesian, for instance, and thought that it was important to include an assessment of the prior probability of a hypothesis, how can this subjective feature be measured?

50. Head, "Extent and Consequences."

51. Simmons et al., "False Positive Psychology," 1362–1363.

52. Simmons et al., "False Positive Psychology," 1365.

53. Novella, "P-Hacking."

54. And, even if it does not, the "bots" are coming. In a recent paper entitled "The Prevalence of Statistical Reporting Errors in Psychology (1985–2013)," M. Nuijten et al. announce the results of a newly created software package called "statcheck" that can be used to check for errors in any APA-style paper. So far, 50,000 papers in psychology have been audited with half showing some sort of mathematical error

(usually tending in the author's favor). These results are published on the PubPeer website (which some have called "methodological terrorism"), but, as the authors point out, the goal is to deter future errors, so there is now a web application that allows authors to check for errors in their own work before their papers are submitted. This too would be a cultural change in science, enabled by better technology. *Behavior Research Methods* 48, no. 4 (2016): 1205–1226, https://mbnuijten.files.wordpress.com/2013/01/nuijtenetal_2015_reportingerrorspsychology1.pdf. See also Brian Resnick, "A Bot Crawled Thousands of Studies Looking for Simple Math Errors: The Results Are Concerning," *Vox*, Sep. 30, 2016, http://www.vox.com/science-and-health/2016/9/30/13077658/statcheck-psychology-replication.

55. Indeed, in some cases peer review is subject to double or even triple-blind reviewing, where in addition to the author not knowing who the reviewer is, the reviewer doesn't know who the author is, and the editor may not know either.

56. John Huizenga, *Cold Fusion: The Scientific Fiasco of the Century* (Rochester: University of Rochester Press, 1992), 235.

57. Huizenga, *Cold Fusion*, 215–236.

58. Huizenga, *Cold Fusion*, 218.

59. Huizenga, *Cold Fusion*, 218.

60. Huizenga, *Cold Fusion*, 57.

61. Gary Taubes, *Bad Science: The Short Life and Weird Times of Cold Fusion* (New York: Random House, 1993).

62. Details of each of these "confirmations" can be found in Taubes, *Bad Science*, where he explores the problems with each: the "Boron" paper was retracted (owing to a discovery that their neutron detector was heat sensitive); the "theoretical" paper about helium was suspect because it had been written expressly to fit the cold fusion findings; and the "excess heat" results in heavy water were flawed by the fact that they also found a reaction in light water, which could be explained by a chemical reaction.

63. Solomon Asch, "Opinions and Social Pressure," *Scientific American* 193, no. 5 (Nov. 1955): 31–35, https://www.panarchy.org/asch/social.pressure.1955.html.

64. Taubes, *Bad Science*, 162.

65. Taubes, *Bad Science*, 162.

66. Wicherts et al., "Willingness to Share Research Data."

67. Fleischmann is guilty of this as well, but it was Pons who repeatedly refused to share data.

68. John Maddox, quoted in Taubes, *Bad Science*, 240.

69. Taubes, *Bad Science*, 191.

70. Taubes, *Bad Science*, 191.

71. Taubes, *Bad Science*, 197.

72. Huizenga, *Cold Fusion*, 234.

73. The fact that there are some people who commit murder does not refute the fact that we live in a society that has laws against murder and attempts to enforce them.

74. Even here, however, this does not necessarily prove bad motive. Perhaps the people who are more scrupulous about keeping their data in good shape so that it can be shared are also those who are most likely to be meticulous in their method.

75. Stuart Firestein, *Failure: Why Science Is So Successful* (Oxford: Oxford University Press, 2015).

76. Firestein, *Failure*.

77. It should go without saying that just as there is a distinction between irreproducibility and fraud, there should be one between retraction and fraud. One should not assume that just because a study has been retracted it is fraudulent. There can be many reasons for retraction, as we shall see.

78. This, however, may be changing—at least in parts of the social sciences—as some new researchers seem eager to make a name for themselves by challenging the work of their more-established colleagues. See Susan Dominus, "When the Revolution Came for Amy Cuddy," *New York Times*, Oct. 18, 2017, https://www.nytimes.com/2017/10/18/magazine/when-the-revolution-came-for-amy-cuddy.html.

79. Ian Sample, "Study Delivers Bleak Verdict on Validity of Psychology Experiment Results," *Guardian*, Aug. 27, 2015, https://www.theguardian.com/science/2015/aug/27/study-delivers-bleak-verdict-on-validity-of-psychology-experiment-results.

80. B. Nosek et al., "Estimating the Reproducibility of Psychological Science," *Science* 349, no. 6251 (Aug. 2015), http://psych.hanover.edu/classes/Cognition/Papers/Science-2015--.pdf. Note that of the 100 studies evaluated, 3 were excluded for statistical reasons, 62 were found to be irreproducible, and only 35 were found to be reproducible.

81. B. Carey, "Many Psychology Findings Not as Strong as Claimed, Study Says," *New York Times*, Aug. 27, 2015.

82. B. Carey, "Psychologists Welcome Analysis Casting Doubt on Their Work," *New York Times*, Aug. 28, 2015.

83. Joel Achenbach, "No, Science's Reproducibility Problem Is Not Limited to Psychology," *Washington Post*, Aug. 28, 2015.

84. Carey, "Many Psychology Findings."

85. A. Nutt, "Errors Riddled 2015 Study Showing Replication Crisis in Psychology Research, Scientists Say," *Washington Post*, March 3, 2016.

86. B. Carey, "New Critique Sees Flaws in Landmark Analysis of Psychology Studies," *New York Times*, March 3, 2016.

87. Nutt, "Errors Riddled 2015 Study."

88. P. Reuell, "Study That Undercut Psych Research Got It Wrong: Widely Reported Analysis That Said Much Research Couldn't Be Reproduced Is Riddled with Its Own Replication Errors, Researchers Say," *Harvard Gazette*, March 3, 2016.

89. Carey, "New Critique Sees Flaws."

90. Reuell, "Study That Undercut Psych Research."

91. Nutt, "Errors Riddled 2015 Study."

92. Reuell, "Study That Undercut Psych Research."

93. Carey, "New Critique Sees Flaws."

94. Carey, "New Critique Sees Flaws."

95. Simonsohn, quoted in Carey, "New Critique Sees Flaws."

96. See http://www.psychologicalscience.org/publications/psychological_science/preregistration.

97. With p-hacking, for instance, can't we imagine a scientist's decision about whether to keep a study open to look for more data as evidence of confirmation bias? Unconsciously, they are rooting for their own theory. Would it necessarily feel wrong to want to gather more data to see if it really worked? Here it seems wise to make note of "Hanlon's razor," which tells us that we should be reluctant to attribute to malice that which can be adequately explained by incompetence.

98. Kevin deLaplante, *The Critical Thinker* (podcast), "Cognitive Biases and the Authority of Science," https://www.youtube.com/watch?v=hZkkY2XVzdw&index=5&list=PLCD69C3C29B645CBC.

99. The task is formidable here, for there are two possible errors: when one waits too long to embrace the truth (as with the reaction to Semmelweis) and when one jumps too soon to a conclusion that outstrips the evidence (as with cold fusion).

6 How the Scientific Attitude Transformed Modern Medicine

1. As we will see in this chapter, though, the birth of modern medicine was beset by numerous setbacks, resistance, and failure to translate understanding into practice, until the scientific attitude was firmly in place.

2. W. Bynum, *The History of Medicine: A Very Short Introduction* (Oxford: Oxford University Press, 2008), 108. See also W. Bynum et al., *The Western Medical Tradition 1800–2000* (Cambridge: Cambridge University Press, 2006), 112.

3. It is important to note, though, that Lister did not "invent" antisepsis, although he is the person most responsible for finding an effective technique and making it routine. See Roy Porter, *The Greatest Benefit to Mankind: A Medical History of Humanity* (New York: Norton, 1999), 370. Even this technique, however, was resisted by some; see Porter, *Greatest Benefit*, 156.

4. Bynum, *History of Medicine*, 91.

5. Though the word *scientist* was not invented until 1833 (by William Whewell), there was of course the earlier Latin word *scientia*, which led to the sixteenth-century use of the word *scientific* that can be roughly translated as "producing knowledge."

6. Bynum, *History of Medicine*, 91. See also Bynum et al., *Western Medical Tradition*, 112.

7. Porter, *Greatest Benefit to Mankind*, 9.

8. Porter, *Greatest Benefit to Mankind*, 57.

9. Porter, *Greatest Benefit to Mankind*, 77.

10. Porter, *Greatest Benefit to Mankind*, 76.

11. Even though some had hoped to grab onto the spirit of scientific advancement that arose as a result of the Enlightenment, "medicine failed to match the achievements of experimental physics or chemistry." Porter, *Greatest Benefit to Mankind*, 248.

12. Porter, *Greatest Benefit to Mankind*, 11.

13. Porter, *Greatest Benefit to Mankind*, 245. It is intriguing to ask why this happened. Porter speculates that "historians have sometimes explained this apparent paradox of Enlightenment medical science—great expectations, disappointing results—as the consequence of over-ambitious theorizing" (248).

14. Porter, *Greatest Benefit to Mankind*, 11, 274. Some, however, will see this as unfair. While it is true that the theory behind vaccines was based not on bench science but on experience with milkmaids and their immunity against cow pox, it was Edward Jenner who embraced the scientific attitude by suggesting "but why think? Why not try the experiment?" (276). Only then, when observation was tested in experiment, did the clinical breakthrough go forward.

15. Porter, *Greatest Benefit to Mankind*, 284, 305.

16. Lewis Thomas, *The Youngest Science: Notes of a Medicine Watcher* (New York: Viking, 1983), 19–20.

17. Ira Rutkow, *Seeking the Cure: A History of Medicine in America* (New York: Scribner, 2010), 37.

18. Rutkow, *Seeking the Cure*, 37–38.

19. Rutkow, *Seeking the Cure*, 44.

20. Cristin O'Keefe Aptowicz, *Dr. Mutter's Marvels: A True Tale of Intrigue and Innovation at the Dawn of Modern Medicine* (New York: Avery, 2014), 31.

21. Porter, *Greatest Benefit to Mankind*, 306.

22. "Being guided by practice and case-oriented, medicine was slow to change. The microscope had been in existence for two hundred years before it became part of everyday medical practice and transformed understanding." Porter, *Greatest Benefit to Mankind*, 525.

23. For anesthesia, some of the holdouts felt that using anesthesia on a woman in labor contradicted the Bible, which said that children should be born in pain as a punishment for original sin. They felt that it was wrong to prevent people from "passing through what God intended them to endure." Rutkow, *Seeking the Cure*, 59–60.

24. Porter, *Greatest Benefit to Mankind*, 431–432.

25. Porter, *Greatest Benefit to Mankind*, 432.

26. Porter, *Greatest Benefit to Mankind*, 433.

27. Rutkow, *Seeking the Cure*, 66.

28. No less a thinker than Rudolf Virchow resisted the germ theory of disease, saying at one point, "If I could live my life over again, I would devote it to proving that germs seek their natural habitat: diseased tissue, rather than being the cause of diseased tissue."

29. Rutkow, *Seeking the Cure*, 79.

30. This quotation comes from John Hughes Bennett, a surgeon and professor in Edinburgh, who continued: "Show them to us, and we shall believe in them. Has anyone seen them yet?" Porter, *Greatest Benefit to Mankind*, 372.

31. Porter, *Greatest Benefit to Mankind*, 436.

32. Porter, *Greatest Benefit to Mankind*, 442.

33. Porter, *Greatest Benefit to Mankind*, 674.

34. Bynum, *Western Medical Tradition*, 112.

35. Porter, *Greatest Benefit to Mankind*, 525.

36. Porter, *Greatest Benefit to Mankind*, 525.

37. Porter, *Greatest Benefit to Mankind*, 527.

38. Porter, *Greatest Benefit to Mankind*, 527.

39. Thomas, *Youngest Science*, 28, 35.

40. James Gleick, *Genius: The Life and Science of Richard Feynman* (New York: Pantheon, 1992), 132.

41. Paul Starr, *The Social Transformation of American Medicine* (New York: Basic Books, 1982).

42. Starr, *Social Transformation of American Medicine*, 39.

43. Starr, *Social Transformation of American Medicine*, 56.

44. Starr, *Social Transformation of American Medicine*, 57.

45. Rutkow, *Seeking the Cure*, 105.

46. "What fundamentally destroyed licensure was the suspicion that it was an expression of favor rather than competence." Starr, *Social Transformation of American Medicine*, 58. Eventually this attitude was swept away by the developments of science (59).

47. Between 1802 and 1876 (which encompassed the era of Jacksonian rebellion against medical societies) sixty-two commercial medical schools opened in the United States. Porter, *Greatest Benefit to Mankind*, 530.

48. Starr, *Social Transformation of American Medicine*, 104.

49. Rutkow, *Seeking the Cure*, 124.

50. Porter, *Greatest Benefit to Mankind*, 530.

51. Porter, *Greatest Benefit to Mankind*, 119.

52. Rutkow, *Seeking the Cure*, 147–148.

53. Porter, *Greatest Benefit to Mankind*, 530–531.

54. Starr, *Social Transformation of American Medicine*, 120–121.

55. Rutkow, *Seeking the Cure*, 164.

56. Bynum, *Western Medical Tradition*, 112.

57. Porter, *Greatest Benefit to Mankind*, 456.

58. Thomas, *Youngest Science*, 35.

59. Porter, *Greatest Benefit to Mankind*, 455–456.

60. There is an excellent account of this story in James Le Fanu, *The Rise and Fall of Modern Medicine* (New York: Carroll and Graf, 1999), 5–15, and also in Porter, *Greatest Benefit to Mankind*, 455–456.

61. Le Fanu, *Rise and Fall of Modern Medicine*, vii.

62. Le Fanu, *Rise and Fall of Modern Medicine*, 5.

63. Le Fanu, *Rise and Fall of Modern Medicine*, 160.

64. Porter, *Greatest Benefit to Mankind*, 455.

65. Le Fanu, *Rise and Fall of Modern Medicine*, 201.

66. Le Fanu, *Rise and Fall of Modern Medicine*, 9.

67. Le Fanu, *Rise and Fall of Modern Medicine*, 10.

68. Porter, *Greatest Benefit to Mankind*, 460.

7 Science Gone Wrong: Fraud and Other Failures

1. Although there are different ways of defining it, the federal standard for research misconduct is intentional "fabrication, falsification, or plagiarism in proposing, performing, or reviewing research, or in reporting research results." Of special note is the further statement that "research misconduct does not include honest error or differences of opinion." See https://www.aps.org/policy/statements/upload/federalpolicy.pdf. This matches most university policies. For a thoughtful history of one university's attempt to come up with its own definition, in conformity with the federal guidelines, see David Goodstein, *On Fact and Fraud: Cautionary Tales from the Front Lines of Science* (Princeton: Princeton University Press, 2010), 67. Of particular note in Goodstein's account is the subtle issue of why it may not be a good thing for a university to have an overly broad definition of prohibited behaviors, nor lump fraud in with all other forms of research misconduct. Caltech's full policy can be found in Goodstein, *On Fact and Fraud*, 136.

2. See text accompanying chapter 2, note 29.

3. Note that as a consequence of this very precise logical definition we have defined not just fraud but also what is *not* fraud. If one has *not* intentionally fabricated or falsified data, then one has *not* committed fraud, and if one has *not* committed fraud, then one has *not* intentionally fabricated or falsified data.

4. And of course it does not follow from thesis (3).

5. In some special cases one probably *could* make a case that p-hacking constituted fraud—for instance, if one knew for a fact that there was no underlying correlation.

6. Naturally, we could change the definition, but see Goodstein, *On Fact and Fraud*, on the problems with including "questionable research practices" in the definition of fraud.

7. Again some of these *might* be fraud, depending on the individual circumstances (see Trivers, "Fraud, Disclosure, and Degrees of Freedom in Science," *Psychology*

Today, May 10, 2012), and there is also the fascinating question of whether things like this can *evolve* into fraud.

8. Goodstein was Caltech's vice provost, who oversaw all cases of scientific misconduct, for nearly twenty years.

9. Goodstein, *On Fact and Fraud*, 2.

10. It is interesting to note that later in his book, Goodstein gives an account of what "self-correction" means that squares nicely with my account of the scientific attitude. He argues that self-correction can hardly mean that we expect individual researchers to question the validity of their own work. Rather, we mean something more like the methods of empirical scrutiny that are implemented by the "scientific community as a whole." Goodstein, *On Fact and Fraud*, 79.

11. "Injecting falsehoods into the body of science is rarely, if ever, the purpose of those who perpetrate fraud. They almost always believe that they are injecting a truth into the scientific record…but without going through all the trouble that the real scientific method demands." Goodstein, *On Fact and Fraud*, 2.

12. Although some philosophers of science have recently taken up the question of fraud, most have not yet graduated from the stereotypical view that all fraud is characterized by "knowingly enter[ing] falsehoods into the information stream of science." See Liam Bright, "On Fraud," *Philosophical Studies* 174, no. 2 (2017): 291–310.

13. See especially Plato, *Theaetetus*, 200d–201c; *Meno*, 86b–c.

14. In his book *The Folly of Fools* (New York: Basic Books, 2011), Robert Trivers offers a stirring discussion of self-delusion, where he offers that perhaps we learn to fool ourselves because it makes us more capable of fooling others.

15. *Meno*, 98b.

16. Perhaps they are worried about litigation, or fear that it will tarnish the reputation of the school where the fraudster worked. Goodstein proposes that one difficulty created by this is that in cases where someone is exonerated they get lots of press, but in cases of actual fraud there is untoward pressure to maintain confidentiality and protect the guilty (*On Fact and Fraud*, xii). (If true, one wonders whether this violates the spirit of science and makes it harder to police the line between mistakes and misconduct.) Goodstein notes, however, one recent case at Stanford where school officials pledged before the investigation was even complete to make the results public (99).

17. In one salient example, Iowa State University scientist Dong-Pyou Han was sentenced to over four years in prison for faking his AIDS research. See Tony Leys, "Ex-Scientist Sentenced to Prison for Academic Fraud," *USA Today*, July 1, 2015, http://www.usatoday.com/story/news/nation/2015/07/01/ex-scientist-sentenced-prison-academic-fraud/29596271/. In another case, South Korean scientist Hwang

Woo-suk was sentenced for fraudulent work on stem cells. See Choe Sang-Hun, "Disgraced Cloning Expert Convicted in South Korea," *New York Times,* Oct. 26, 2009, https://www.nytimes.com/2009/10/27/world/asia/27clone.html.

18. Though this can happen. Consider the case of Thereza Imanishi-Kari, who was investigated by the federal government and charged with falsifying laboratory data; the case collapsed and she was exonerated. "The Fraud Case That Evaporated," *New York Times* (Opinion), June 25, 1996, http://www.nytimes.com/1996/06/25/opinion/the-fraud-case-that-evaporated.html.

19. See Carolyn Y. Johnson, "Ex-Harvard Scientist Fabricated, Manipulated Data, Report Says, *Boston Globe,* Sept. 5, 2012, https://www.bostonglobe.com/news/science/2012/09/05/harvard-professor-who-resigned-fabricated-manipulated-data-says/6gDVkzPNxv1ZDkh4wVnKhO/story.html.

20. Goodstein, *On Fact and Fraud,* 65.

21. Goodstein, *On Fact and Fraud,* 60–61.

22. Once again, consider Trivers's work on the possible connection between deception and self-deception.

23. Robert Park, *Voodoo Science: The Road from Foolishness to Fraud* (Oxford: Oxford University Press, 2000), 10.

24. Goodstein, *On Fact and Fraud,* 129.

25. Goodstein, *On Fact and Fraud,* 70.

26. Is there a possible analogy here between fraud and pseudoscience? Is self-delusion what we condemn in pseudoscientists? See text accompanying chapter 5, note 53, where I claim that the difference between techniques used by scientists and pseudoscientists is only a matter of degree.

27. Of course, it might depend on why the data were not stored correctly. Deleting one's original data sets is an extremely bad practice, but if it is done to cover up an inquiry into irregularities in one's research, it can constitute fraud.

28. Goodstein, *On Fact and Fraud,* xiii–xiv.

29. Some might argue, though, that this sort of preoccupation with the question of "intent" could have the opposite effect and make the distinction between good and bad science even more impenetrable. Yet do we not already do this in cases of fraud? Sometimes intent has to be inferred from action, but this is no excuse for focusing only on a proxy, like article retraction.

30. Seth Mnookin, *The Panic Virus: The True Story Behind the Vaccine–Autism Controversy* (New York: Simon and Schuster, 2011), 109.

31. Michael Specter, *Denialism* (New York: Penguin, 2009), 71.

32. Mnookin, *Panic Virus*, 236.

33. Mnookin, *Panic Virus*, 305.

34. Mnookin, *Panic Virus*, 19.

35. Some were getting the word out: Jennifer Steinhauser, "Rising Public Health Risk Seen as More Parents Reject Vaccines," *New York Times*, March 21, 2008.

36. Brian Deer, "How the Case against the MMR Vaccine Was Fixed," *British Medical Journal* 342 (2011): c5347.

37. Deer, "How the Case against the MMR Vaccine Was Fixed," c5347.

38. F. Godlee et al., "Wakefield Article Linking MMR Vaccine and Autism Was Fraudulent," *British Medical Journal* 342 (2011): c7452.

39. Godlee et al., "Wakefield Article," 2.

40. D. K. Flaherty, "The Vaccine–Autism Connection: A Public Health Crisis Caused by Unethical Medical Practices and Fraudulent Science," *Annals of Pharmacotherapy* 45, no. 10 (2011): 1302–1304.

41. Mark Berman, "More Than 100 Confirmed Cases of Measles in the U.S., CDC Says," *Washington Post*, Feb. 2, 2015.

42. All of his coauthors claimed not to have known of Wakefield's deep conflict of interest or manipulation of data, though two of them were later investigated by the General Medical Council in Britain, and one was found guilty of misconduct.

43. This has been exacerbated by continued publicity for Wakefield's discredited hypothesis by public figures such as Robert F. Kennedy, Jr., *Thimerosal: Let the Science Speak: The Evidence Supporting the Immediate Removal of Mercury—a Known Neurotoxin—from Vaccines* (New York: Skyhorse Publishing, 2015), and actor Robert De Niro's 2016 decision to screen (and then pull) the film *Vaxxed: From Cover-Up to Conspiracy* at the Tribeca Film Festival. De Niro later said that he regretted pulling the film.

44. Michael D. Lemonick, "When Scientists Screw Up," *Science*, Oct. 15, 2002.

8 Science Gone Sideways: Denialists, Pseudoscientists, and Other Charlatans

1. Some may prefer the term "denier" over "denialist" for specific beliefs (e.g., vaccine denier) but the phenomenon itself is called "denialism." Given this, I think it is clearer to call people who engage in this practice "denialists," which has become accepted usage at least since Michael Specter's *Denialism: How Irrational Thinking Hinders Scientific Progress, Harms the Planet, and Threatens Our Lives* (New York: Penguin, 2009).

2. I will pursue examples of this later in the chapter.

3. And it can cause great harm and suffering. Although it is much less commonly discussed than climate change, AIDS denial is a particularly pernicious example. Between 2000 and 2004, President Thabo Mbeki's governmental policy of refusing Western antiretroviral drugs, which was informed by maverick scientist Peter Duesberg from Berkeley, California, is estimated by researchers from Harvard University to have caused 300,000 avoidable deaths in South Africa. See Sarah Boseley, "Mbeki AIDS Denial 'Caused 300,000 Deaths,'" *Guardian*, Nov. 26, 2008, https://www.theguardian.com/world/2008/nov/26/aids-south-africa.

4. See "Public Praises Science; Scientists Fault Public, Media," Pew Research Center, *U.S. Politics & Policy*, July 9, 2009, http://www.people-press.org/2009/07/09/public-praises-science-scientists-fault-public-media/; Cary Funk and Brian Kennedy, "Public Confidence in Scientists Has Remained Stable for Decades," Pew Research Center, April 6, 2017, http://www.pewresearch.org/fact-tank/2017/04/06/public-confidence-in-scientists-has-remained-stable-for-decades/; The National Science Foundation, "Science and Engineering Indicators 2014," http://www.nsf.gov/statistics/seind14/index.cfm/chapter-7/c7h.htm.

5. See Solomon Asch's experimental work on social conformity from the 1950s ("Opinions and Social Pressure," *Scientific American* 193, no. 5 [Nov. 1955]: 31–35), which shows not only that agreement solidifies belief, but that dissonance with one's peers can cause subjects to *change* their beliefs to obvious falsehoods.

6. Lee McIntyre, *Respecting Truth: Willful Ignorance in the Internet Age* (New York: Routledge, 2015).

7. Noretta Koertge, "Belief Buddies versus Critical Communities," in *Philosophy of Pseudoscience*, ed. M. Pigliucci and M. Boudry (Chicago: University of Chicago Press, 2013), 169.

8. Carl Sagan, *The Demon-Haunted World: Science as a Candle in the Dark* (New York: Ballantine Books, 1996).

9. Sagan, *Demon-Haunted World*, 304. See also his discussion on 31 and 305–306.

10. Sagan, *Demon-Haunted World*, 305.

11. Sagan, *Demon-Haunted World*, 305.

12. Sagan, *Demon-Haunted World*, 13, 100.

13. Sagan lists over seventy different examples of pseudoscience in his book, *Demon-Haunted World*, 221–222.

14. Sagan, *Demon-Haunted World*, 187. "Keeping an open mind is a virtue ... but not so open your brains fall out."

15. To his credit, Sagan lists three claims in the field of ESP research that he thinks deserve investigation. Sagan, *Demon-Haunted World*, 302. One of these—the claim

that thought alone can influence a random number generator—will be considered later in this chapter. But one should not conclude from this that Sagan is a pushover, for he also endorses Laplace's insight: "Extraordinary claims require extraordinary evidence."

16. See http://www.csicop.org/about/csicop.

17. Later in this chapter, I will examine in more depth the sense in which scientists are skeptical but denialists are not. It might be better to say that denialists are "selective" but, as we shall see, the criterion for selectivity in denialism (which cherry picks data for the sake of consistency with its preferred ideology) is scientifically illegitimate.

18. Sagan, *Demon-Haunted World*, 304.

19. Here I am obviously going beyond what Sagan actually said—for he did not address denialism—but I nonetheless find it useful to use this matrix as a device to consider the similarities and differences between denialism and pseudoscience.

20. Note here the distinction between Andrew Wakefield himself, who fraudulently claimed that vaccines cause autism, versus the denialism of those who continue to believe his claim even after his fraud has been exposed.

21. Of course, they have *beliefs*; they are just not scientific beliefs. They are ideological. Thus it matters what denialists take to be the standard for their other beliefs.

22. This might seem curious. If their beliefs are ideological, why do they think that they need any evidence? Because they are not willing to admit that their beliefs are ideological. But then they are trapped in a double standard, for once we are playing the evidence game how can they hold that their own views are better warranted than those of science? This just seems delusional. Once a denialist insists that you must disprove his evidence, while dismissing your own out of hand, it seems best just to walk away.

23. We might usefully compare scientific skepticism with what Robert Merton called "organized skepticism" in his essay "The Normative Structure of Science" (1942), reprinted as chapter 13 in *The Sociology of Science*, ed. Robert Merton (Chicago: University of Chicago Press, 1973).

24. Though there is now a fledgling movement called "experimental philosophy."

25. Although such a statement may seem uncomfortably close for string theorists, they would surely welcome opportunities to put their theories to the test, even if such opportunities are not available at the moment.

26. I discuss this further in my article "The Price of Denialism," *New York Times*, Nov. 11, 2015, https://opinionator.blogs.nytimes.com/2015/11/07/the-rules-of-denialism/.

27. In one provocative study, Brendan Nyhan and Jason Reifler showed that exposing those with biased beliefs to evidence that their beliefs were wrong could actually

have a "backfire effect." See Nyhan and Reifler, "When Corrections Fail: The Persistence of Political Misperceptions, *Political Behavior* 22, no. 2 (2010): 303–330, https://www.dartmouth.edu/~nyhan/nyhan-reifler.pdf. For an earlier study that paved the way for this result, see C. Lord, L. Ross, and M. Lepper, "Biased Assimilation and Attitude Polarization: The Effects of Prior Theories on Subsequently Considered Evidence," *Journal of Personality and Social Psychology* 37, no. 11 (Nov. 1979): 2098–2109, http://dx.doi.org/10.1037/0022-3514.37.11.2098.

28. It is a fair point that Sagan did not present this as such, but it seems a natural extension of his views. I maintain that Sagan's matrix is also wrong about pseudoscientists being open to new ideas; see below in this chapter.

29. For instance, President Mbeki's health minister claimed that AIDS could be cured with garlic and lemon juice. See Celia W. Dugger, "Study Cites Toll of AIDS Policy in South Africa," *New York Times*, Nov. 25, 2008, https://www.nytimes.com/2008/11/26/world/africa/26aids.html.

30. Massimo Pigliucci, *Nonsense on Stilts: How to Tell Science from Bunk* (Chicago: University of Chicago Press, 2010), 137.

31. In a 2004 literature review published in *Science*, historian of science Naomi Oreskes found that of 928 peer-reviewed scientific papers on global climate change published between 1993 and 2003, exactly *zero* of them disagreed with the idea that human-caused global warming was occurring. A follow-up review in 2012 found that of 13,950 peer-reviewed papers on climate change from 1991 to 2012, only 0.17 percent rejected global warming. For discussion of some of the scientific evidence see http://climate.nasa.gov/evidence/ and http://www.ucsusa.org/our-work/global-warming/science-and-impacts/global-warming-science#.V-beXvkrK1s.

32. Naomi Oreskes and Erik Conway, *Merchants of Doubt* (New York: Bloomsbury, 2010).

33. McIntyre, *Respecting Truth*, 72–80.

34. For instance, Sen. James Inhofe, Sen. Rick Santorum, and President Donald Trump.

35. Rebecca Kaplan and Ellen Uchimiya, "Where the 2016 Republican Candidates Stand on Climate Change," CBSNews.com, Sept. 1, 2015, http://www.cbsnews.com/news/where-the-2016-republican-candidates-stand-on-climate-change/.

36. Thomas R. Karl et al., "Poissible Artifacts of Data Biases in the Recent Global Surface Warming Hiatus," *Science*, June 26, 2015, http://science.sciencemag.org/content/348/6242/1469.full.

37. "Scientific Evidence Doesn't Support Global Warming, Sen. Ted Cruz Says," NPR, Dec. 9, 2015, http://www.npr.org/2015/12/09/459026242/scientific-evidence-doesn-t-support-global-warming-sen-ted-cruz-says.

38. Justin Gillis, "Global Warming 'Hiatus' Challenged by NOAA Research," *New York Times*, June 4, 2015, http://www.nytimes.com/2015/06/05/science/noaa-research-presents-evidence-against-a-global-warming-hiatus.html.

39. Michele Berger, "Climate Change Not on Hiatus, New Research Shows," Weather.com, June 4, 2015, https://weather.com/science/environment/news/no-climate-change-hiatus-noaa-says.

40. For the uncorrected graph, see http://scienceblogs.com/significantfigures/files/2013/04/updated-global-temperature.png. For the corrected graph, see http://cdn.arstechnica.net/wp-content/uploads/2015/06/noaa_karl_etal-640x486.jpg.

41. The Associated Press recently made no one happy with a policy change to stop calling climate change denialists either "deniers" or "skeptics," preferring the term "doubter." On one side, they had complaints that the term "denier" was too closely associated with "Holocaust denier," and on the other those who said that the term "skeptic" was and should be associated with scientific skepticism. Their compromise, of course, leaves the impression that there is still "room for doubt" about the truth of climate change, which is arguably no better than using the term "skeptic." Puneet Kollipara, "At Associated Press, No More Climate Skeptics or Deniers," Sciencemag.org, Sep. 23, 2015, http://www.sciencemag.org/news/2015/09/associated-press-no-more-climate-skeptics-or-deniers.

42. Yes they still exist, and they have a gorgeous website. One wonders, though, whether they believe that any of their Internet traffic involves satellites, for what would they orbit? http://www.theflatearthsociety.org/home/index.php.

43. McIntyre, *Respecting Truth*, 73. In an intriguing follow up, however, it was found that virtually *all* of the contrarian papers on climate change had methodological flaws in them. R. E. Benestad, D. Nuccitelli, S. Lewandowsky, et al., "Learning from Mistakes in Climate Research," *Theoretical and Applied Climatology* 126, nos. 3–4 (2016): 699, https://doi.org/10.1007/s00704-015-1597-5.

44. Taken from a 2009 Pew Research poll cited in https://ncse.com/blog/2013/08/how-many-creationists-science-0014996.

45. Thomas Kuhn explains this well in his *The Structure of Scientific Revolutions* (Chicago: University of Chicago Press, 1962).

46. See Koertge, "Belief Buddies versus Critical Communities."

47. Philip Bump, "Ted Cruz Compares Climate Change Activists to 'Flat-Earthers': Where to Begin?" *Washington Post*, March 25, 2015, https://www.washingtonpost.com/news/the-fix/wp/2015/03/25/ted-cruz-compares-climate-change-activists-to-flat-earthers-where-to-begin/. Though, technically speaking, what Galileo disputed was geocentrism; not all geocentrists believed in a flat Earth.

48. John Soennichsen, *Bretz's Flood: The Remarkable Story of a Rebel Geologist and the World's Greatest Flood* (Seattle: Sasquatch Books, 2008), 126.

49. Soennichsen, *Bretz's Flood*, 131.

50. Soennichsen, *Bretz's Flood*, 133.

51. Soennichsen, *Bretz's Flood*, 143–144.

52. Many also, no doubt, would compare Bretz to his contemporary geologist Alfred Wegener, whose theory of plate tectonics was the subject of enormous ridicule and rejection, with vindication coming only many years after Wegener's death. Soennichsen, *Bretz's Flood*, 165–168.

53. Most of Bretz's principle critics had never visited the scablands. Soennichsen, *Bretz's Flood*, 201. This reminds one of those who rejected Galileo's theory, yet refused to look through his telescope.

54. Soennichsen, *Bretz's Flood*, 160.

55. Soennichsen, *Bretz's Flood*, 191, 207.

56. http://magazine.uchicago.edu/0912/features/legacy.shtml.

57. Soennichsen, *Bretz's Flood*, 144.

58. Soennichsen, *Bretz's Flood*, 226. This is how science sometimes works; see Kuhn, *Structure of Scientific Revolutions*.

59. Soennichsen, *Bretz's Flood*, 228.

60. Soennichsen, *Bretz's Flood*,. 231.

61. For example: http://www.godsaidmansaid.com/printtopic.asp?ItemId=1354.

62. This is not to say that Bretz undermined uniformitarianism in general. Most geological features can in fact be accounted for by gradual change over a vast period of time. What he did, however, was show that there were exceptions to it as an explanation for *all* geological phenomena, in particular for the features of the scabland region.

63. Note that this is the point of my earlier claim that Pigliucci seems wrong when he says that science is "what scientists do," for what if they behave like the ones who fought Bretz?

64. Note for instance Stephen Jay Gould's theory of punctuated equilibrium, which, while scientifically controversial, nonetheless demonstrates that it is possible to offer a nontheological hypothesis to propose a sudden change in natural events: http://www.pbs.org/wgbh/evolution/library/03/5/l_035_01.html.

65. Another example might be seen in some of the mistakes made by Stephen Jay Gould in his work on scientific racism. See Robert Trivers's blistering indictment of Gould, whom he felt committed borderline fraud in service to his political agenda: "Fraud in the Imputation of Fraud: The Mis-Measure of Stephen Jay Gould," *Psychology Today*, Oct. 4, 2012, https://www.psychologytoday.com/blog/the-folly-fools/201210/fraud-in-the-imputation-fraud.

66. While it is true that other considerations like internal consistency, explanatory power, scope, fruitfulness, and the like may sometimes be relevant to theory choice in science, ideally these are used within a context in which more than one theory fits the evidence and a choice must still be made. This is to say that fit with the evidence is a necessary condition that pseudoscience does not meet. See Kuhn, *Structure of Scientific Revolutions*, for further discussion of the idea that "social factors" can sometimes be crucial for theory choice in science. Yet even Kuhn believes that if a theory does not fit with the evidence, no amount of other considerations should be able to rescue it.

67. Accurate figures are hard to come by, but this estimate is from Paul Kurtz, the prominent American philosopher and skeptic, from a 1985 study of the astrology industry. Brian Lehrer, "Born Under a Dollar Sign Astrology is Big Business, Even If It Is All Taurus," *Orlando Sentinel*, Nov. 10, 1985, http://articles.orlandosentinel.com/1985-11-10/news/0340290056_1_astrology-columns-un-sign-astrology-scientific-fact.

68. See "34 Billion Spent Yearly on Alternative Medicine," NBCNews.com, July 30, 2009, http://www.nbcnews.com/id/32219873/ns/health-alternative_medicine/t/billion-spent-yearly-alternative-medicine/#.V-bqYPkrK1t.

69. Whether one chooses to call pseudoscience "fake" or "fraud" perhaps matters little, but I prefer to think of those who embrace the scientific attitude and then cheat on it as frauds, whereas those who only pretend to embrace it while claiming to do scientific work are fakes.

70. My favorite example here, courtesy of Carl Sagan, is why a spirit medium who claims to be in touch with the ghost of Fermat never asks him to provide the details of his missing proof.

71. For those who are curious about these ongoing efforts, one might take a look at *The Skeptical Inquirer*, which is published by CSICOP http://www.csicop.org/si. Sagan's work is masterful here. See in particular his chapter "The Fine Art of Baloney Detection," in *Demon-Haunted World*, 203–218.

72. Michael Ruse, *But Is It Science? The Philosophical Question in the Creation/Evolution Controversy* (Amherst, NY: Prometheus Books, 1996); Pigliucci, *Nonsense on Stilts*, 160–186; McIntyre, *Dark Ages: The Case for a Science of Human Behavior* (Cambridge, MA: MIT Press, 2006), 85–92; McIntyre, *Respecting Truth*, 64–71.

73. In contradistinction to the popular myth of a "moral victory" at trial (which was made into the play and movie *Inherit the Wind*), Tennessee's antievolution law stayed on the books until it was repealed in 1967.

74. Ruse, *But Is It Science?* 320.

75. McIntyre, *Respecting Truth*, 67–68.

76. https://www.nytimes.com/2001/04/08/us/darwin-vs-design-evolutionists-new-battle.html. It might have helped the intelligent design proponents more if one of

their central texts had not been a word-for-word revision of an earlier creationist text, with the only change being removal of the word "creationist" and substitution of the word "design proponent," except for one place they missed which said "cdesign proponentsists."

77. See https://ncse.com/library-resource/discovery-institutes-model-academic-freedom-statute.

78. McIntyre, *Respecting Truth*, 69.

79. Indeed, in an absolutely brilliant paper, Nicholas Matzke has done a phylogenetic analysis of the family relationship between all of these bills, using one of the primary tools of evolution to examine antievolution legislation. "The Evolution of Antievolution Policies after *Kitzmiller v. Dover*," *Science* 351, no. 6268 (Jan. 1, 2016): 28–30. http://science.sciencemag.org/content/early/2015/12/16/science.aad4057.

80. Laura Moser, "Another Year, Another Anti-Evolution Bill in Oklahoma," *Slate*, January 25, 2016. http://www.slate.com/blogs/schooled/2016/01/25/oklahoma_evolution_controversy_two_new_bills_present_alternatives_to_evolution.html

81. John Timmer, "This Year's First Batch of Anti-Science Education Bills Surface in Oklahoma,"*Ars Technica*, Jan. 24, 2016, http://arstechnica.com/science/2016/01/this-years-first-batch-of-anti-science-education-bills-surface-in-oklahoma/.

82. https://ncse.com/creationism/general/chronology-academic-freedom-bills.

83. Lee McIntyre, "The Attack on Truth," *Chronicle of Higher Education*, June 8, 2015, http://www.chronicle.com/article/The-Attack-on-Truth/230631.

84. http://www.evolutionnews.org/2015/06/willful_ignoran096781.html; http://www.evolutionnews.org/2015/06/say_what_you_wa096811.html.

85. "After *Kitzmiller*, even the Discovery Institute, the institutional home of ID Theory, claimed it had never encouraged teaching ID in public schools (incorrectly) and heavily promoted 'Academic Freedom Acts' (AFAs), aimed at encouraging teachers to promote antievolutionism." Matzke, "Evolution of Antievolution Policies," 1.

86. The classic source is Michael Ruse, *But Is It Science?* See also Massimo Pigliucci, *Denying Evolution: Creationism, Scientism, and the Nature of Science* (Sunderland, MA: Sinauer Associates, 2002); Donald Prothero, *Evolution: What the Fossils Say and Why It Matters* (New York: Columbia University Press, 2007); Sahotra Sarkar, *Doubting Darwin? Creationist Designs on Evolution* (New York: Wiley-Blackwell, 2007).

87. The decision in the Kitzmiller trial is fascinating reading and can be found at https://ncse.com/files/pub/legal/kitzmiller/highlights/2005-12-20_Kitzmiller_decision.pdf. Echoing Judge Overton's decision in *McLean v. Arkansas* in 1981, Judge Jones held that if a theory cannot be proven wrong by any possible evidence, then

it is not scientific. The full transcript of the decision in the McLean decision can be found in Ruse, *But Is It Science?*

88. See Ruse, *But Is It Science?* See also Pigliucci, *Denying Evolution*; Prothero, *Evolution*; Sarkar, *Doubting Darwin?*

89. Bobby Henderson, *The Gospel of the Flying Spaghetti Monster* (New York: Villard Books, 2006).

90. An ID theorist might complain that it is unfair to compare their theory to phlogiston because that was refuted, but the problem is that their own theory is unfalsifiable, so it offers no possibility to *be* refuted.

91. See http://skepdic.com/pear.html.

92. Pigliucci, *Nonsense on Stilts*, 78.

93. Pigliucci, *Nonsense on Stilts*, 77–80. On p. 78, Pigliucci argues that what PEAR was doing was not pseudoscience, it was just wrong.

94. Indeed, some might argue that the best evidence that the RNGs were not truly random was the effect itself. This of course begs the question against the research and we must try to do better than this.

95. Robert Park, *Voodoo Science: The Road from Foolishness to Fraud* (Oxford: Oxford University Press, 2000), 199.

96. http://www.csicop.org/si/show/pear_lab_closes_ending_decades_of_psychic_research.

97. Benedict Carey, "A Princeton Lab on ESP Plans to Close Its Doors," *New York Times*, Feb. 10, 2007.

98. http://skepdic.com/pear.html.

99. Carey, "A Princeton Lab."

9 The Case for the Social Sciences

1. This is at least in part due to the problem of *reflexive prediction*, which is when a human subject takes a prediction of their behavior into account, which in turn affects their actions. Note, however, that this does not necessarily make human behavior unpredictable at either the individual or the group level; but it does make it impossible to run a counterfactual experiment and see what would have happened if you hadn't shared any information with the subject in the first place.

2. Lee McIntyre, *Laws and Explanation in the Social Sciences: Defending a Science of Human Behavior* (Boulder: Westview Press, 1996); Lee McIntyre, *Dark Ages: The Case for a Science of Human Behavior* (Cambridge, MA: MIT Press, 2006).

3. Of course, attitude can influence method (as I acknowledged in *Dark Ages*, 20), but I now wish to place more emphasis on the importance of that dynamic.

4. McIntyre, *Dark Ages*, 93.

5. Popper, "Prediction and Prophecy in the Social Sciences," in *Conjectures and Refutations: The Growth of Scientific Knowledge* (New York: Harper Torchbooks, 1965), 336–346.

6. McIntyre, *Dark Ages*, 123, note 4.

7. McIntyre, *Respecting Truth: Willful Ignorance in the Internet Age* (New York: Routledge, 2015), 37.

8. See Steven Yoder, "Life on the List," *American Prospect*, April 4, 2011, http://prospect.org/article/life-list. In this case, perhaps there are normative considerations as well: we might believe that sex offenders just deserve to be punished beyond their sentence or that neighbors have a right to know where they live. If that is the case, though, the recidivism rate is irrelevant.

9. McIntyre, *Dark Ages*, 63–68.

10. Robert Trivers, "Fraud, Disclosure, and Degrees of Freedom in Science," *Psychology Today* (blog entry: May 10, 2012), https://www.psychologytoday.com/blog/the-folly-fools/201205/fraud-disclosure-and-degrees-freedom-in-science.

11. Jay Gabler and Jason Kaufman, "Chess, Cheerleading, Chopin: What Gets You Into College?" *Contexts* 5, no. 2 (Spring 2006): 45–49. If one digs, one finds that this was based on an earlier study in which the researchers did control for socioeconomic status, but even so it seems suspect to pursue the idea that there is a *direct causal link* between parental museum visitation and college admission. For instance, they speculate that name dropping about the latest art exhibit at the Whitney Museum during an interview might make an applicant seem more like "college material." This hypothesis, however, was never tested. Jason Kaufman and Jay Gabler, "Cultural Capital and the Extracurricular Activities of Girls and Boys in the College Attainment Process," *Poetics* 32 (2004): 145–168.

12. Emile Durkheim, *The Rules of Sociological Method*, author's preface (Paris, 1895).

13. Think of medicine before Semmelweis: arguments about intuition with no data to check against are dangerous in scientific inquiry. We need to keep ourselves honest by looking for empirical evidence, wherever it is available.

14. One long-standing study has offered a superior, cost-effective alternative to simultaneous lineups, but it has been roundly rejected out of hand by numerous law enforcement agencies, including the FBI. R. C. Lindsay and G. L. Wells, "Improving Eyewitness Identification from Lineups: Simultaneous versus Sequential Lineup Presentation," *Journal of Applied Psychology* 70 (1985): 556–564.

15. Susan Fiske and Cydney Dupree, "Gaining Trust as Well as Respect in Communicating to Motivated Audiences about Science Topics," *Proceedings of the National Academy of Science* 111, suppl. 4 (Sept. 16, 2014): 13593–13597, http://www.pnas.org/content/111/Supplement_4/13593.full.

16. Fiske and Dupree, "Gaining Trust."

17. M. Brewer and R. Brown, "Intergroup Relations," in *Handbook of Social Psychology*, ed. D. Gilbert, S. Fiske, and G. Lindzey (New York: Oxford University Press, 1998), 554–595.

18. They do have an argument linking whether someone is "on my side" with whether they are judged to be "warm." What they need, however, is some evidence that there is a two-way relationship between being warm and being trustworthy.

19. Fiske and Dupree, "Gaining Trust."

20. This happens more often in social scientific research than one might think. Even the most beautifully set up experiment can fail if the wrong conclusion is drawn from the evidence. Many social scientific experiments show *something*, just perhaps not what their researchers claim.

21. Fiske and Dupree, "Gaining Trust."

22. Fiske and Dupree, "Gaining Trust."

23. There are surely many more examples that could be cited here. But am I now myself subject to the criticism that I have just cherry picked a particularly egregious example? I don't think so, for this study was conducted by researchers at an R1 research university (Princeton) and published in a prestigious journal. As for selection bias, I came across this study as I was looking for information about how scientists might become better communicators about issues such as climate change.

24. For more on the classic assumptions that economists and others have made about human rationality, and how they break down in the face of experimental evidence, see Daniel Kahneman, *Thinking Fast and Slow* (New York: Farrar, Straus and Giroux, 2011).

25. One very early challenge to this, however, can be found in Herbert Simon's work on bounded rationality and satisficing. See Herbert A. Simon, "Theories of Bounded Rationality," in *Decision and Organization*, ed. C. B. McGuire and Roy Radner (Amsterdam: North-Holland, 1972).

26. Note that speculation is sometimes necessary even in the best-conducted scientific studies, but here it was centered around a tight research question, under controlled conditions, that enabled follow up in further scientific work.

27. Sheena Iyengar and Mark Lepper, "Rethinking the Value of Choice: A Cultural Perspective on Intrinsic Motivation," *Journal of Personality and Social Psychology* 76, no. 3 (1999): 349–366.

28. McIntyre, *Respecting Truth*, 29–36.

29. Fiske and Dupree, "Gaining Trust."

10 Valuing Science

1. And there is certainly a difference from saying that it is both necessary *and* sufficient. See chapter 4.

2. Recall Laudan: philosophers largely abandoned this project for thirty years after his 1983 essay.

3. And when they do, we must investigate further. Just imagine if the cold fusion theorists had had a little more success with their predictions.

Bibliography

Achenbach, Joel. "No, Science's Reproducibility Problem Is Not Limited to Psychology." *Washington Post*, Aug. 28, 2015.

Achinstein, Peter. *The Book of Evidence*. Oxford: Oxford University Press, 2003.

Achinstein, Peter, ed. *Scientific Evidence: Philosophical Theories and Applications*. Baltimore: Johns Hopkins University Press, 2005.

Achinstein, Peter, and Stephen Barker. *The Legacy of Logical Positivism*. Baltimore: Johns Hopkins University Press, 1969.

Alcock, James. "Back from the Future: Parapsychology and the Bem Affair." *Skeptical Inquirer*, Jan. 6, 2011, http://www.csicop.org/specialarticles/show/back_from_the_future.

Aptowicz, Cristin O'Keeffe. *Dr. Mutter's Marvels*. New York: Avery, 2015.

Asch, Solomon. "Opinions and Social Pressure." *Scientific American* 193, no. 5 (Nov. 1955): 31–35.

Aschwanden, Christie. "Science Isn't Broken." *FiveThirtyEight*, Aug. 19, 2015, https://fivethirtyeight.com/features/science-isnt-broken/.

Ayer, A. J. *Language, Truth, and Logic*. New York: Dover, 1955.

Bacon, Francis. *The New Atlantis*. 1627.

Bacon, Francis. *The New Organon*. 1620.

Baker, Vic. "Megafloods of the Ice Age." *Nova*, Sept. 20, 2005, http://www.pbs.org/wgbh/nova/earth/megafloods-of-the-ice-age.html.

Berman, Mark. "More Than 100 Confirmed Cases of Measles in the U.S." *Washington Post*, Feb. 2, 2015, https://www.washingtonpost.com/news/to-your-health/wp/2015/02/02/more-than-100-confirmed-cases-of-measles-in-the-u-s-cdc-says/?utm_term=.4a8312361f3b.

Blachowicz, James. "There Is No Scientific Method." *New York Times*, July 4, 2016.

Bokulich, Alisa. "Pluto and the 'Planet Problem': Folk Concepts and Natural Kinds in Astronomy." *Perspectives on Science* 22, no. 4 (2014): 464–490.

Boseley, Sarah. "Mbeki Aids Denial 'Caused 300,000 Deaths.'" *Guardian*, Nov. 26, 2008.

Boudry, Maarten. "Loki's Wager and Laudan's Error." In *The Philosophy of Pseudoscience: Reconsidering the Demarcation Problem*, ed. Massimo Pigliucci and Maarten Boudry, 79–98. Chicago: University of Chicago Press, 2013.

Boudry, Maarten. "Plus Ultra: Why Science Does Not Have Limits." In *Science Unlimited: The Challenges of Scientism*, ed. Maarten Boudry and Massimo Pigliucci. Chicago: University of Chicago Press, 2017.

Boudry, Maarten, and Massimo Pigliucci. "Science Unlimited? The Challenges of Scientism." In *Science Unlimited: The Challenges of Scientism*, ed. Maarten Boudry and Massimo Pigliucci. Chicago: University of Chicago Press, 2017.

Brainard, Jeffrey. "As U.S. Releases New Rules on Scientific Fraud, Scholars Debate How Much and Why It Occurs." *Chronicle of Higher Education*, Dec. 8, 2000.

Bright, Liam. "On Fraud." *Philosophical Studies* 174, no. 2 (2017): 291–310.

Brower, Kenneth. "The Danger of Cosmic Genius." *Atlantic*, Dec. 2010.

Brown, James Robert. *Who Rules in Science? An Opinionated Guide to the Wars*. Cambridge, MA: Harvard University Press, 2004.

Buchanan, Mark. "Stop Hiding from the Truth." *Bloomberg News*, July 6, 2016, https://www.bloomberg.com/view/articles/2015-07-06/stop-hiding-from-the-truth.

Bynum, William. *The History of Medicine: A Very Short Introduction*. Oxford: Oxford University Press, 2008.

Bynum, William, Anne Hardy, Stephen Jacyna, Christopher Lawrence, and E. M. (Tilli) Tansey. *The Western Medical Tradition 1800 to 2000*. Cambridge: Cambridge University Press, 2006.

Carey, Benedict. "Many Psychology Findings Not as Strong as Claimed, Study Says." *New York Times*, Aug. 27, 2015.

Carey, Benedict. "New Critique Sees Flaws in Landmark Analysis of Psychology Studies." *New York Times*, March 3, 2016.

Carey, Benedict. "A Princeton Lab on ESP Plans to Close Its Doors." *New York Times*, Feb. 10, 2007.

Carey, Benedict. "Psychologists Welcome Analysis Casting Doubt on Their Work." *New York Times*, Aug. 28, 2015.

Carey, Bjorn. "Stanford Researchers Uncover Patterns in How Scientists Lie about Data." *Stanford News*, Nov. 16, 2015.

Carroll, Robert, "The Princeton Engineering Anomalies Research (PEAR)." In *The Skeptic's Dictionary*, accessed Aug. 28, 2018, http://skepdic.com/pear.html.

Cartwright, Nancy. *The Dappled World: A Study of the Boundaries of Science*. Cambridge: Cambridge University Press, 1999.

Castelvecchi. Davide. "Is String Theory Science?" *Scientific American*, Dec. 23, 2015, https://www.scientificamerican.com/article/is-string-theory-science/.

Chalmers, Alan. *What Is This Thing Called Science?* Indianapolis: Hackett, 2013.

Chang, Hasok. *Is Water H_2O? Evidence, Realism and Pluralism*. Dordrecht: Springer, 2014.

Cioffi, Frank. "Pseudoscience: The Case of Freud's Sexual Etiology of the Neuroses." In *The Philosophy of Pseudoscience: Reconsidering the Demarcation Problem*, ed. Massimo Pigliucci and Maarten Boudry, 321–340. Chicago: University of Chicago Press, 2013.

Coffman, E. J. "Warrant without Truth?" *Synthese* 162 (2008): 173–194.

Coll, Steve. *Private Empire: ExxonMobil and American Power*. New York: Penguin, 2013.

Curd, Martin. "Philosophy of Pseudoscience: Reconsidering the Demarcation Problem." *Notre Dame Philosophical Reviews*, July 17, 2014, http://ndpr.nd.edu/news/philosophy-of-pseudoscience-reconsidering-the-demarcation-problem/.

Davis, Rebecca. "The Doctor Who Championed Hand-Washing and Briefly Saved Lives." NPR, Jan. 12, 2015, http://www.npr.org/sections/health-shots/2015/01/12/375663920/the-doctor-who-championed-hand-washing-and-saved-women-s-lives.

Dawid, Richard. *String Theory and the Scientific Method*. Cambridge: Cambridge University Press, 2014.

Dawkins, Richard. *Climbing Mount Improbable*. New York: Norton, 2016.

Deer, Brian. "British Doctor Who Kicked-Off Vaccines-Autism Scare May Have Lied, Newspaper Says." *Los Angeles Times*, Feb. 9, 2009.

Deer, Brian. "How the Case against the MMR Vaccine Was Fixed." *British Medical Journal* 342 (2011): case 5347.

Department of Health and Human Services. "Findings of Research Misconduct: Notice Number: NOT-OD-12-149." Sept. 10, 2012. https://grants.nih.gov/grants/guide/notice-files/NOT-OD-12-149.html.

Devlin, Hannah. "Cut-Throat Academia Leads to 'Natural Selection of Bad Science,' Claims Study." *Guardian*, Sept. 21, 2016.

Duhem, Pierre. *The Aim and Structure of Physical Theory*. Princeton: Princeton University Press, 1991.

Dupre, John. *The Disorder of Things: Metaphysical Foundations of the Disunity of Science*. Cambridge, MA: Harvard University Press, 1995.

Durkheim, Emile. *The Rules of Sociological Method*. Paris, 1895.

Dyson, Freeman. "Misunderstandings, Questionable Beliefs Mar Paris Climate Talks." *Boston Globe*, Dec. 3, 2015.

Elgin, Catherine. *True Enough*. Cambridge, MA: MIT Press, 2017.

Ellis, Paul. "Why Are Journal Editors Increasingly Asking Authors to Report Effect Sizes?" *Effectsizefaq.com*, May 31, 2010, https://effectsizefaq.com/2010/05/31/why-are-journal-editors-increasingly-asking-authors-to-report-effect-sizes/.

Fairweather, Abrol, and Linda Zagzebski, eds. *Virtue Epistemology: Essays on Epistemic Virtue and Responsibility*. Oxford: Oxford University Press, 2011.

Fairweather, Abrol. *Virtue Epistemology Naturalized: Bridges between Virtue Epistemology and Philosophy of Science*. Dordrecht: Springer, 2015.

Feleppa, Robert. "Kuhn, Popper, and the Normative Problem of Demarcation." In *Philosophy of Science and the Occult*, ed. Patrick Grim. Albany: SUNY Press, 1990.

Feyerabend, Paul. *Against Method*. London: Verso, 1975.

Feynman, Richard. *Surely You're Joking, Mr. Feynman*. New York: Norton, 1997.

Firestein, Stuart. *Failure: Why Science Is So Successful*. Oxford: Oxford University Press, 2016.

Firestein, Stuart. "When Failing Equals Success." *Los Angeles Times*, Feb. 14, 2016.

Fiske, Susan, and Cydney Dupree. "Gaining Trust as Well as Respect in Communicating to Motivated Audiences about Science Topics." *Proceedings of the National Academy of Sciences* 111, suppl. 4 (2014).

Flaherty, DK. "The Vaccine-Autism Connection: A Public Health Crisis Caused by Unethical Medical Practices and Fraudulent Science." *Annals of Pharmacotherapy* 45, no. 10 (Oct. 2011): 1302–1304.

Foran, Clare. "Ted Cruz Turns Up the Heat on Climate Change." *Atlantic*, Dec. 9, 2015.

Fuller, Steve. *Philosophy of Science and Its Discontents*. New York: Guilford Press, 1992.

Gabler, Jay, and Jason Kaufman. "Chess, Cheerleading, Chopin: What Gets You into College?" *Contexts* 5, no. 2 (2006): 45–49.

Galison, Peter, and Bruce Hevly, eds. *Big Science: The Growth of Large-Scale Research*. Palo Alto: Stanford University Press, 1992.

Giere, Ronald. *Understanding Scientific Reasoning*. New York: Harcourt, 1991.

Gilbert, Daniel, Gary King, Stephen Pettigrew, and Timothy Wilson. "Comment on 'Estimating the Reproducibility of Psychological Science.'" *Science* 351, no. 6277 (2016). http://science.sciencemag.org/content/351/6277/1037.2.

Gillis, Justin. "Scientists Warn of Perilous Climate Shift within Decades, Not Centuries." *New York Times*, March 22, 2016.

Gleick, James. *Genius: The Life and Science of Richard Feynman*. New York: Vintage, 1993.

Glymour, Clark. *Theory and Evidence*. Princeton: Princeton University Press, 1980.

Godlee, F. "Wakefield Article Linking MMR Vaccine and Autism Was Fraudulent." *British Medical Journal* 342 (2011): case 7452.

Goodman, Nelson. *Fact, Fiction, and Forecast*. Cambridge, MA: Harvard University Press, 1955.

Goodstein, David. *On Fact and Fraud: Cautionary Tales from the Front Lines of Science*. Princeton: Princeton University Press, 2010.

Greco, John, and John Turri, eds. *Virtue Epistemology: Contemporary Readings*. Cambridge, MA: MIT Press, 2012.

Greene, Brian. *The Elegant Universe*. New York: Norton, 2010.

Guijosa, Alberto. "What Is String Theory?" accessed Aug. 28, 2018, http://www.nuclecu.unam.mx/~alberto/physics/string.html.

Haack, Susan. *Defending Science—within Reason: Between Scientism And Cynicism*. Amherst, NY: Prometheus, 2007.

Haack, Susan. *Evidence and Inquiry: A Pragmatist Reconstruction of Epistemology*. Amherst, NY: Prometheus, 2009.

Haack, Susan. "Six Signs of Scientism." *Skeptical Inquirer* 37, no. 6 (Nov.–Dec. 2013); 38, no. 1 (Jan.–Feb. 2014).

Hacking, Ian. *The Social Construction of What?* Cambridge, MA: Harvard University Press, 1999.

Hall, Shannon. "Exxon Knew about Climate Change almost 40 Years Ago." *Scientific American*, Oct. 26, 2015.

Hansen, James. *Storms of My Grandchildren*. New York: Bloomsbury, 2010.

Hansson, Sven Ove. "Defining Pseudoscience and Science." In *The Philosophy of Pseudoscience: Reconsidering the Demarcation Problem*, ed. Massimo Pigliucci and Maarten Boudry, 61–77. Chicago: University of Chicago Press, 2013.

Hansson, Sven Ove. "Science and Pseudo-Science." In *The Stanford Encyclopedia of Philosophy*, last modified April 11, 2017, https://plato.stanford.edu/entries/pseudo-science/.

Harris, Paul. "Four US States Considering Laws That Challenge Teaching of Evolution." *Guardian*, Jan. 31, 2013.

Hartgerink, C. H. J. "Renanalyzing Head et al. (2015): No Widespread P-Hacking After All?" *Authorea*, Sept. 12, 2016, https://www.authorea.com/users/2013/articles/31568.

Head, Meagan, Luke Holman, Rob Lanfear, Andrew Kahn, and Michael Jennions. "The Extent and Consequences of P-Hacking in Science." *PLOS Biology*, March 13, 2015, http://journals.plos.org/plosbiology/article?id=10.1371/journal.pbio.1002106.

Hempel, Carl. *Philosophy of Natural Science*. New York: Prentice Hall, 1966.

Henderson. Bobby. *The Gospel of the Flying Spaghetti Monster*. New York: Villard Books, 2006.

Hicks, Daniel, and Thomas Stapleford. "The Virtues of Scientific Practice: MacIntyre, Virtue Ethics, and the Historiography of Science." *Isis* 107, no. 3 (Sept. 2016): 449–472.

Holton, Gerald. *Science and Anti-Science*. Cambridge, MA: Harvard University Press, 1998.

Holton, Gerald. *The Scientific Imagination*. Cambridge, MA: Harvard University Press, 1998.

Hoyningen-Huene, Paul. *Systematicity: The Nature of Science*. Oxford: Oxford University Press, 2013.

Huber, Rose. "Scientists Seen as Competent But Not Trusted by Americans." Princeton University, Woodrow Wilson School of Public and International Affairs, Sept. 22, 2014.

Huizenga, John. *Cold Fusion: The Scientific Fiasco of the Century*. Rochester: University of Rochester Press, 1992.

Hull, David. *Science as a Process: An Evolutionary Account of the Social and Conceptual Development of Science*. Chicago: University of Chicago Press, 1990.

Hume, David. *An Enquiry Concerning Human Understanding*. London, 1748.

Hume, David. *A Treatise of Human Nature*. London, 1740.

Insel, Thomas. "Director's Blog: P-Hacking." The National Institute of Mental Health, Nov. 14, 2014, https://www.nimh.nih.gov/about/directors/thomas-insel/blog/2014/p-hacking.shtml.

Ioannidis, John. "Why Most Published Research Findings Are False." *PLOS Medicine*, Aug. 30, 2005, http://journals.plos.org/plosmedicine/article?id=10.1371/journal.pmed.0020124.

Iyengar, Sheena, and Mark Lepper. "Rethinking the Value of Choice: A Cultural Perspective on Intrinsic Motivation." *Journal of Personality and Social Psychology* 76, no. 3 (1999): 349–366.

Iyengar, Sheena. *The Art of Choosing*. New York: Twelve, 2010.

Jaschik, Scott. "What Really Counts in Getting In." *Inside Higher Education*, May 31, 2006.

Jeffers, Stanley. "PEAR Lab Closes, Ending Decades of Psychic Research." *Skeptical Inquirer* 31, no. 3 (May–June 2007).

Johnson, Carolyn. "Former Harvard professor Marc Hauser Fabricated, Manipulated Data, US Says." *Boston Globe*, Sept. 5, 2012.

Johnson, Carolyn. "Harvard Report Shines Light on Ex-Researcher's Misconduct." *Boston Globe*, May 30, 2014.

Kahn, Brian. "No Pause in Global Warming." *Scientific American*, June 4, 2015.

Kahneman, Daniel. *Thinking Fast and Slow*. New York: Farrar, Straus and Giroux, 2013.

Karl, Thomas, Anthony Arguez, Boyin Huang, Jay H. Lawrimore, James R. McMahon, Matthew J. Menne, et al. "Possible Artifacts of Data Biases in the Recent Global Surface Warming Hiatus." *Science* 348, no. 6242 (June 26, 2015): 1469–1472.

Kaufman, Jason, and Jay Gabler. "Cultural Capital and the Extracurricular Activities of Girls and Boys in the College Attainment Process." *Poetics* 32 (2004): 145–168.

King, Gary, Robert Keohane, and Sidney Verba. *Designing Social Inquiry: Scientific Inference in Qualitative Research*. Princeton: Princeton University Press, 1994.

Kitcher, Philip. *The Advancement of Science: Science without Legend, Objectivity without Illusions*. Oxford: Oxford University Press, 1995.

Kitcher, Philip. *Science in a Democratic Society*. Amherst: Prometheus Books, 2011.

Kitcher, Philip. *Science, Truth, and Democracy*. Oxford: Oxford University Press, 2003.

Kluger, Jeffrey. "Modern Science Has a Publish-or-Perish Problem." *Time*, Aug. 20, 2015, 19.

Koertge, Noretta, ed. *A House Built on Sand: Exposing Postmodernist Myths about Science*. New York: Oxford University Press, 2000.

Koertge, Noretta. "Belief Buddies versus Critical Communities." In *The Philosophy of Pseudoscience: Reconsidering the Demarcation Problem*, ed. Massimo Pigliucci and Maarten Boudry, 165–180. Chicago: University of Chicago Press, 2013.

Koertge, Noretta. "A Bouquet of Values." In *Scientific Values and Civic Virtues*, ed. Noretta Koertge, 9–24. Oxford: Oxford University Press, 2005.

Koertge, Noretta, ed. *Scientific Values and Civic Virtues*. Oxford: Oxford University Press, 2005.

Kuhn, Thomas. "Logic of Discovery or Psychology of Research?" In *The Philosophy of Karl Popper*, ed. P. A. Schilpp, 798–819. LaSalle: Open Court, 1974.

Kuhn, Thomas. "Objectivity, Value Judgment, and Theory Choice." In *The Essential Tension: Selected Studies in Scientific Tradition and Change*, 320–339. Chicago: University of Chicago Press, 1974.

Kuhn, Thomas. *The Road since Structure*. Chicago: University of Chicago Press, 2000.

Kuhn, Thomas. *The Structure of Scientific Revolutions*. Chicago: University of Chicago Press, 1962.

Ladyman, James. "Toward a Demarcation of Science from Pseudoscience." In *The Philosophy of Pseudoscience: Reconsidering the Demarcation Problem*, ed. Massimo Pigliucci and Maarten Boudry, 45–59. Chicago: University of Chicago Press, 2013.

Lakatos, Imre, and Alan Musgrave, eds. *Criticism and the Growth of Knowledge*. Cambridge: Cambridge University Press, 1970.

Lange, Marc. *Natural Laws in Scientific Practice*. Oxford: Oxford University Press, 2000.

Laudan, Larry. *Beyond Positivism and Relativism: Theory, Method, and Evidence*. Boulder: Westview Press, 1996.

Laudan, Larry. "The Demise of the Demarcation Problem." In Laudan, *Beyond Positivism and Relativism*, 210–222. Boulder: Westview Press, 1996.

Laudan, Larry. *Progress and Its Problems: Towards a Theory of Scientific Growth*. Berkeley: University of California Press, 1977.

Laudan, Larry. "Science at the Bar—Causes for Concern." In *Beyond Positivism and Relativism: Theory, Method, and Evidence*, 223–230. Boulder: Westview Press, 1996.

Le Fanu, James. *The Rise and Fall of Modern Medicine*. New York: Carroll and Graf, 1999.

Lemonick. Michael. "When Scientists Screw Up." *Slate*, Oct. 15, 2012, http://www.slate.com/articles/health_and_science/science/2012/10/scientists_make_mistakes_how_astronomers_and_biologists_correct_the_record.html.

Longino, Helen. "Science and the Common Good: Thoughts on Philip Kitcher's *Science, Truth, and Democracy*." *Philosophy of Science* 69 (Dec. 2002): 560–568.

Longino, Helen. *Science as Social Knowledge: Values and Objectivity in Scientific Inquiry*. Princeton: Princeton University Press, 1990.

Longino, Helen. "Underdetermination: A Dirty Little Secret?" *STS Occasional Papers 4*. London: Department of Science and Technology Studies, University College London, 2016.

Lynch, Michael. *True to Life: Why Truth Matters*. Cambridge, MA: MIT Press, 2004.

Machamer, Peter, Marcello Pera, and Aristides Baltas, eds. *Scientific Controversies: Philosophical and Historical Perspectives*. Oxford: Oxford University Press, 2000.

MacIntyre, Alasdair. *After Virtue: A Study in Moral Theory*. 2nd ed. South Bend: University of Notre Dame Press, 1984.

Mahner, Martin. "Science and Pseudoscience." In *The Philosophy of Pseudoscience: Reconsidering the Demarcation Problem*, ed. Massimo Pigliucci and Maarten Boudry, 29–43. Chicago: University of Chicago Press, 2013.

Marcus, Adam, and Oransky, Ivan. "The Lessons of Famous Science Frauds." *Verge*, June 9, 2015.

Marcus, Adam, and Oransky, Ivan. "What's Behind Big Science Frauds?" *New York Times*, May 22, 2015.

Matzke, Nicholas. "The Evolution of Antievolution Policies after *Kitzmiller v. Dover*." *Science* 351, no. 6268 (Jan. 1, 2016): 28–30.

Maxwell, Nicholas. "The Rationality of Scientific Discovery, Part I." *Philosophy of Science* 41, no. 2 (1974): 123–153.

Maxwell, Nicholas. "The Rationality of Scientific Discovery, Part II." *Philosophy of Science* 41, no. 3 (1974): 247–295.

Mayo, Deborah. "Ducks, Rabbits, and Normal Science: Recasting the Kuhn's-Eye View of Popper." *British Journal for the Philosophy of Science* 47, no. 2 (1996): 271–290.

Mayo, Deborah. *Error and the Growth of Experimental Knowledge*. Chicago: University of Chicago Press, 1996.

Mayo, Deborah. "Popper on Pseudoscience: A Comment on Pigliucci." *Errorstatistics.com*, Sept. 16, 2015, https://errorstatistics.com/2015/09/16/popper-on-pseudoscience-a-comment-on-pigliucci-i/.

Mayo, Deborah, and Aris Spanos, eds. *Error and Inference: Recent Exchanges on Experimental Reasoning, Reliability, Objectivity, and Rationality*. New York: Cambridge University Press, 2010.

McIntyre, Lee. "Accommodation, Prediction, and Confirmation." *Perspectives on Science* 9, no. 3 (2001): 308–323.

McIntyre, Lee. *Dark Ages: The Case for a Science of Human Behavior*. Cambridge, MA: MIT Press, 2006.

McIntyre, Lee. "Intentionality, Pluralism and Redescription." *Philosophy of the Social Sciences* 34, no. 4 (2004): 493–505.

McIntyre, Lee. *Laws and Explanation in the Social Sciences: Defending a Science of Human Behavior*. Boulder: Westview Press, 1996.

McIntyre, Lee. "The Price of Denialism." *New York Times*, Nov. 7, 2015.

McIntyre, Lee. "Redescription and Descriptivism." *Behavior and Philosophy* 32, no. 2 (2004): 453–464.

McIntyre, Lee. *Respecting Truth: Willful Ignorance in the Internet Age*. New York: Routledge, 2015.

McIntyre, Lee. "The Scientific Attitude toward Explanation." In *Modes of Explanation: Affordances for Action and Prediction*, ed. Michael Lissack and Abraham Graber, 229–232. New York: Palgrave, 2014.

McMullin, Ernan. "Values in Science." *PSA: Proceedings of the Biennial Meeting of the Philosophy of Science Association* 4 (1982): 3–28.

Meikle, James, and Sarah Boseley. "MMR Row Doctor Andrew Wakefield Struck Off Register." *Guardian*, May 24, 2010.

Mellor, D. H. "The Warrant of Induction." In *Matters of Metaphysics*. Cambridge: Cambridge University Press, 1991.

Mercier, Hugo, and Dan Sperber. "Why Do Humans Reason? Arguments for an Argumentative Theory." *Behavioral and Brain Sciences* 34, no. 2 (2011): 57–111.

Merton, Robert. "The Normative Structure of Science." In *The Sociology of Science*, ed. Robert Merton, chapter 13. Chicago: University of Chicago Press, 1973. (Originally published 1942.)

Mirowski, Philip. *More Heat than Light: Economics as Social Physics*. Cambridge: Cambridge University Press, 1990.

Mnookin, Seth. *The Panic Virus: The True Story Behind the Vaccine-Autism Controversy*. New York: Simon and Schuster, 2011.

Moser, Laura. "Another Year, Another Anti-Evolution Bill in Oklahoma." *Slate.com* (blog), Jan. 25, 2016, http://www.slate.com/blogs/schooled/2016/01/25/oklahoma_evolution_controversy_two_new_bills_present_alternatives_to_evolution.html.

Muntersbjorn, Madeline. "Francis Bacon's Philosophy of Science: Machina Intellectus and Forma Indita." *Philosophy of Science* 70, no. 5 (Dec. 2003): 1137–1148.

Nagel, Ernest. *The Structure of Science*. New York: Routledge, 1961.

Nahigyan, Pierce. "Global Warming Never Stopped." *Huffington Post*, Dec. 3, 2015.

Nickles, Thomas. "The Problem of Demarcation: History and Future." In *The Philosophy of Pseudoscience: Reconsidering the Demarcation Problem*, ed. Massimo Pigliucci and Maarten Boudry, 101–120. Chicago: University of Chicago Press, 2013.

Nosek, Brian. "Estimating the Reproducibility of Psychological Science." *Science* 349, no. 6251 (Aug. 28, 2015): aac4716.

Novella, Steven. "P-Hacking and Other Statistical Sins." *Neurologica*, Feb. 13, 2014, http://theness.com/neurologicablog/index.php/p-hacking-and-other-statistical-sins/.

Novella, Steven. "Publishing False Positives." *Neurologica*, Jan. 5, 2012, http://theness.com/neurologicablog/index.php/publishing-false-positives/.

NPR News. "Scientific Evidence Doesn't Support Global Warming, Sen. Ted Cruz Says." Dec. 9, 2015, http://www.npr.org/2015/12/09/459026242/scientific-evidence-doesn-t-support-global-warming-sen-ted-cruz-says.

Nuccitelli, Dana. "Here's What Happens When You Try to Replicate Climate Contrarian Papers." *Guardian*, Aug. 25, 2015.

Nuijten, M., C. H. Hartgerink, M. A. van Assen, S. Epskamp, and J. M. Wicherts. "The Prevalence of Statistical Reporting Errors in Psychology (1985–2013)." *Behavioral Research* 48, no. 4 (Oct. 23, 2015), https://www.ncbi.nlm.nih.gov/pubmed/26497820.

Nutt, Amy. "Errors Riddled 2015 Study Showing Replication Crisis in Psychology Research, Scientists Say." *Washington Post*, March 3, 2016.

Nuzzo, Regina. "Scientific Method: Statistical Errors." *Nature* 506, no. 7487 (Feb. 12, 2014), http://www.nature.com/news/scientific-method-statistical-errors-1.14700.

Nyhan, Brendan, and Jason Reifler. "When Corrections Fail: The Persistence of Political Misperceptions." *Political Behavior* 32 (2010): 303–330.

Okasha, Samir. *Philosophy of Science: A Very Short Introduction*. Oxford: Oxford University Press, 2002.

Oreskes, Naomi, and Erik Conway. *Merchants of Doubt: How a Handful of Scientists Obscured the Truth on Issues from Tobacco Smoke to Global Warming*. New York: Bloomsbury Press, 2010.

Pappas, Stephanie. "Climate Change Disbelief Rises in America." *LiveScience*, Jan. 16, 2014, http://www.livescience.com/42633-climate-change-disbelief-rises.html.

Park, Robert. *Superstition: Belief in the Age of Science*. Princeton: Princeton University Press, 2008.

Park, Robert. *Voodoo Science: The Road from Foolishness to Fraud*. Oxford: Oxford University Press, 2000.

Passmore, John. *Science and Its Critics*. New Brunswick: Rutgers University Press, 1978.

Peirce, Charles. "The Scientific Attitude and Fallibilism." In *The Collected Papers of Charles S. Peirce*, vol. 1, *Principles of Philosophy*. Cambridge, MA: Harvard University Press, 1931.

Pigliucci, Massimo. "The Demaration Problem: A (Belated) Response to Laudan." In *The Philosophy of Pseudoscience: Reconsidering the Demarcation Problem*, ed. Massimo Pigliucci and Maarten Boudry, 9–28. Chicago: University of Chicago Press, 2013.

Pigliucci, Massimo. *Denying Evolution*. Sunderland, MA: Sinauer, 2002.

Pigliucci, Massimo. *Nonsense on Stilts: How to Tell Science from Bunk*. Chicago: University of Chicago Press, 2010.

Pigliucci, Massimo. "Scientism and Pseudoscience: In Defense of Demarcation Projects." In *Science Unlimited*. Chicago: University of Chicago Press, 2017.

Pigliucci, Massimo, and Maarten Boudry. "Introduction: Why the Demarcation Problem Matters." In *The Philosophy of Pseudoscience: Reconsidering the Demarcation Problem*, ed. Massimo Pigliucci and Maarten Boudry, 1–6. Chicago: University of Chicago Press, 2013.

Pigliucci, Massimo, and Maarten Boudry, eds. *The Philosophy of Pseudoscience: Reconsidering the Demarcation Problem*. Chicago: University of Chicago Press, 2013.

Plait, Phil. "Scientists Explain Why Ted Cruz Is Wrong about the Climate." *Mother Jones*, Jan. 19, 2016, https://www.motherjones.com/environment/2016/01/ted-cruz-satellite-date-climate-change/.

Popper, Karl. *The Logic of Scientific Discovery*. New York: Routledge, 1959.

Popper, Karl. "Natural Selection and the Emergence of Mind." *Dialectica* 32, nos. 3–4 (1978): 339–355.

Popper, Karl. "Normal Science and Its Dangers." In *Problems in the Philosophy of Science*, ed. Imre Lakatos and Alan Musgrave, 51–58. Cambridge: Cambridge University Press, 1970.

Popper, Karl. "On the Non-Existence of Scientific Method." In *Realism and the Aim of Science*. Lanham, MD: Rowman and Littlefield, 1983.

Popper, Karl. *The Open Universe*. New York: Routledge, 1992.

Popper, Karl. "Prediction and Prophecy in the Social Sciences." In *Conjectures and Refutations*, 336–346. New York: Harper Torchbooks, 1965.

Popper, Karl. "Remarks on the Problems of Demarcation and of Rationality." In *Problems in the Philosophy of Science*, ed. Imre Lakatos and Alan Musgrave, 88–102. Amsterdam: North-Holland, 1968.

Popper, Karl. "Replies to My Critics." In *The Philosophy of Karl Popper*, ed. P. A. Schilpp. LaSalle: Open Court, 1974.

Popper, Karl. "Science: Conjectures and Refutations." In *Conjectures and Refutations*, 33–65. New York: Harper Torchbooks, 1965.

Popper, Karl. *Unended Quest*. LaSalle: Open Court, 1982.

Porter, Roy. *The Greatest Benefit to Mankind: A Medical History of Humanity*. New York: Norton, 1997.

Prothero, Donald. *Reality Check: How Science Deniers Threaten Our Future*. Bloomington: Indiana University Press, 2013.

Putnam, Hilary. "The 'Corroboration' of Theories." In *The Philosophy of Karl Popper*, ed. P. A. Schilpp, 221–240. LaSalle: Open Court, 1974.

Putnam, Hilary. *Meaning and the Moral Sciences*. London: Routledge, 1978.

Quine, W. V. O. "Two Dogmas of Empiricism." *Philosophical Review* 60 (1951): 20–43.

Reardon, Sara. "Uneven Response to Scientific Fraud." *Nature* 523 (July 9, 2015): 138–139.

Resnick, Brian. "Study: Elite Scientists Can Hold Back Science." *Vox*, Dec. 15, 2015, http://www.vox.com/science-and-health/2015/12/15/10219330/elite-scientists-hold-back-progress.

Resnick, Brian. "A Bot Crawled Thousands of Studies Looking for Simple Math Errors: The Results Are Concerning." *Vox*, Sept. 30, 2016, http://www.vox.com/science-and-health/2016/9/30/13077658/statcheck-psychology-replication.

Reuell, Peter. "Study That Undercut Psych Research Got It Wrong." *Harvard Gazette*, March 2, 2016, http://news.harvard.edu/gazette/story/2016/03/study-that-undercut-psych-research-got-it-wrong/.

Rosenberg, Alex. *The Atheist's Guide to Reality: Enjoying Life without Illusions*. New York: Norton, 2011.

Rouse, Joseph, *How Scientific Practices Matter: Reclaiming Philosophical Naturalism*. Chicago: University of Chicago Press, 2003.

Ruse, Michael, ed. *But Is It Science?* Amherst: Prometheus Books, 1996.

Ruse, Michael. "Evolution: From Pseudoscience to Popular Science, from Popular Science to Professional Science." In *The Philosophy of Pseudoscience: Reconsidering the Demarcation Problem*, ed. Massimo Pigliucci and Maarten Boudry, 225–244. Chicago: University of Chicago Press, 2013.

Ruse, Michael. "Karl Popper's Philosophy of Biology." In *But Is It Science?* ed. Michael Ruse. Amherst: Prometheus Books, 1996.

Rutkow, Ira. *Seeking the Cure: A History of Medicine in America*. New York: Scribner, 2010.

Sagan, Carl. *The Demon Haunted World: Science as a Candle in the Dark*. New York: Random House, 1995.

Salmon Wesley. "Hans Reichenbach's Vindication of Induction." *Erkenntnis* 35, no. 1 (1991): 99–122.

Sargent, Rose-Mary. "Virtues and the Scientific Revolution." In *Scientific Values and Civic Virtues*, ed. N. Koertge, 71–80. Oxford: Oxford University Press, 2005.

Scerri, Eric. *A Tale of Seven Scientists and a New Philosophy of Science*. Oxford: Oxford University Press, 2016.

Scerri, Eric. "Has Chemistry Been at Least Approximately Reduced to Quantum Mechanics?" In *PSA 1994*, vol. 1, ed. D. Hull, M. Forbes, and R. Burian, 160–170. East Lansing, MI: Philosophy of Science Association, 1994.

Scerri, Eric, and Lee McIntyre. "The Case for the Philosophy of Chemistry." *Synthese* 111, no. 3 (1997): 213–232.

Schilpp, Paul, ed. *The Philosophy of Karl Popper*. LaSalle: Open Court, 1974.

Settle, Thomas. "The Rationality of Science versus the Rationality of Magic." *Philosophy of the Social Sciences* 1 (1971): 173–194.

Shapin, Steven. *The Scientific Life: A Moral History of a Late Modern Vocation*. Chicago: University of Chicago Press, 2010.

Shapin, Steven. "Trust, Honesty, and the Authority of Science." In *Society's Choices: Social and Ethical Decision Making in Biomedicine*, ed. R. Bulger, E. Bobby, and H. Fineberg, 388–408. Washington, DC: National Academy Press, 1995.

Shepphard, Kate. "Ted Cruz: 'Global Warming Alarmists Are the Equivalent of the Flat-Earthers.'" *Huffington Post*, March 25, 2015.

Shermer, Michael. *The Believing Brain*. New York: Times Books, 2011.

Shermer, Michael. "Science and Pseudoscience." In *The Philosophy of Pseudoscience: Reconsidering the Demarcation Problem*, ed. Massimo Pigliucci and Maarten Boudry, 203–223. Chicago: University of Chicago Press, 2013.

Siegel, Ethan. "Why String Theory Is Not a Scientific Theory." *Forbes*, Dec. 23, 2015.

Simmons, Joseph, Leif Nelson, and Uri Simonsohn. "False-Positive Psychology: Undisclosed Flexibility in Data Collection and Analysis Allows Presenting Anything as Significant." *Psychological Science* 22, no. 11 (2011): 1359–1366.

Smolin, Lee. *The Trouble with Physics: The Rise of String Theory, the Fall of a Science, and What Comes Next*. New York: Mariner Books, 2007.

Soennichsen, John. *Bretz's Flood: The Remarkable Story of a Rebel Geologist and the World's Greatest Flood*. Seattle: Sasquatch Books, 2008.

Soennichsen, John. "Legacy: J Harlen Bretz (1882–1981)." *University of Chicago Magazine*, http://magazine.uchicago.edu/0912/features/legacy.shtml.

Solomon, Miriam. *Social Empiricism*. Cambridge, MA: MIT Press, 2001.

Sosa, Ernest. "The Raft and the Pyramid: Coherence versus Foundations in the Theory of Knowledge." *Midwest Studies in Philosophy* 5 (1980): 3–25.

Specter, Michael. *Denialism*. New York: Penguin, 2009.

Stanford, Kyle. *Exceeding Our Grasp*. Oxford: Oxford University Press, 2006.

Starr, Paul. *The Social Transformation of American Medicine*. New York: Basic Books, 1982.

Steinhauser, Jennifer. "Rising Public Health Risk Seen as More Parents Reject Vaccines." *New York Times*, March 21, 2008.

Stemwedel, Janet. "Drawing the Line between Science and Pseudo-Science." *Scientific American*, Oct. 4, 2011, https://blogs.scientificamerican.com/doing-good-science/drawing-the-line-between-science-and-pseudo-science/.

Sullivan, Gail, and Richard Feinn. "Using Effect Size—or Why the P Value Is Not Enough." *Journal of Graduate Medical Education* 4, no. 3 (Sept. 2012), https://www.ncbi.nlm.nih.gov/pmc/articles/PMC3444174/.

Sunstein, Cass. *Infotopia: How Many Minds Produce Knowledge*. Oxford: Oxford University Press, 2006.

Taubes, Gary. *Bad Science: The Short Life and Weird Times of Cold Fusion*. New York: Random House, 1993.

Thaler, Richard, and Cass Sunstein. *Nudge*. New Haven: Yale University Press, 2008.

Thomas, Lewis. *The Youngest Science: Notes of a Medicine Watcher*. New York: Viking, 1983.

Timmer, John. "This Year's First Batch of Anti-Science Education Bills Surface in Oklahoma." *Arts Technica*, Jan. 24, 2016, https://arstechnica.com/science/2016/01/this-years-first-batch-of-anti-science-education-bills-surface-in-oklahoma/.

Trivers, Robert. *The Folly of Fools: the Logic of Deceit and Self-Deception in Human Life*. New York: Basic Books, 2011.

Trivers, Robert. "Fraud, Disclosure, and Degrees of Freedom in Science." *Psychology Today*, May 10, 2012, https://www.psychologytoday.com/blog/the-folly-fools/201205/fraud-disclosure-and-degrees-freedom-in-science.

Trivers, Robert. "Fraud in the Imputation of Fraud: The Mis-measure of Stephen Jay Gould." *Psychology Today*, Oct. 4, 2012, https://www.psychologytoday.com/blog/the-folly-fools/201210/fraud-in-the-imputation-fraud.

Turri, John, Mark Alfano, and John Greco. "Virtue Epistemology."In *The Stanford Encyclopedia of Philosophy* (Summer 2017 Edition), ed. Edward N. Zalta, https://plato.stanford.edu/archives/sum2017/entries/epistemology-virtue/.

Union of Concerned Scientists, The. "The Climate Deception Dossiers." July 2015, http://www.ucsusa.org/sites/default/files/attach/2015/07/The-Climate-Deception-Dossiers.pdf.

van Fraassen, Bas. *The Scientific Image*. Oxford: Clarendon Press, 1980.

Wade, Nicholas. "Harvard Finds Scientist Guilty of Misconduct." *New York Times*, Aug. 20, 2010.

Waller, John. *Fabulous Science: Fact and Fiction in the History of Scientific Discovery*. Oxford: Oxford University Press, 2002.

Wicherts, Jelte, Marjan Bakker, and Dylan Moelnaar. "Willingness to Share Research Data Is Related to the Strength of the Evidence and the Quality of Reporting of Statistical Results." *PLOS ONE* 6, no. 11 (Nov. 2011).

Wigner, Eugene. "The Unreasonable Effectiveness of Mathematics in the Natural Sciences." *Communication in Pure and Applied Mathematics* 13, no. 1 (Feb. 1960).

Williams, Bernard. *Truth and Truthfulness: An Essay in Genealogy*. Princeton: Princeton University Press, 2002.

Williams, Richard, and Daniel Robinson, eds. *Scientism: The New Orthodoxy*. New York: Bloomsbury Academic, 2016.

Woit, Peter. "Is String Theory Even Wrong?" *American Scientist*, March–April 2002, http://www.americanscientist.org/issues/pub/is-string-theory-even-wrong/.

Woit, Peter. *Not Even Wrong: The Failure of String Theory and the Search for Unity in Physical Law*. New York: Basic Books, 2007.

Woit, Peter. "String Theory and the Scientific Method." May 14, 2013, http://www.math.columbia.edu/~woit/wordpress/?p=5880.

Wolchover, Natalie. "Physicists and Philosophers Hold Peace Talks." *Atlantic*, Dec. 22, 2015, https://www.theatlantic.com/science/archive/2015/12/physics-philosophy-string-theory/421569/.

Wray, K. Brad. *Kuhn's Evolutionary Social Epistemology*. Cambridge: Cambridge University Press, 2011.

Wray, K. Brad. "Kuhn's Social Epistemology and the Sociology of Science." In *Kuhn's Structure of Scientific Revolutions—50 Years On*, ed. W. Devlin and A. Bokulich, 167–183. Boston Studies in the Philosophy of Science, Vol. 311. New York: Springer, 2015.Index

Index